THE Parent Connection for SINGAPORE MATH

Tools to Help Them "Get It" & Get Behind It

by Sandra Chen

Crystal Springs
BOOKS
A division of SDE Staff Development for EDUCATORS.

Peterborough, New Hampshire

Dedication

To my husband, for understanding my insane work ethic and supporting me throughout all of my endeavors.

Published by Crystal Springs Books
A division of Staff Development for Educators (SDE)
10 Sharon Road, PO Box 500
Peterborough, NH 03458
1-800-321-0401
www.SDE.com/crystalsprings

© 2008 Sandra Chen
Illustrations © 2008 Crystal Springs Books

Published 2008
Printed in the United States of America
12 11 10   2 3 4 5

ISBN: 978-1-934026-05-2

Library of Congress Cataloging-in-Publication Data

Chen, Sandra, 1981-
  The parent connection for Singapore math : tools to help them get it
& get behind it / by Sandra Chen.
      p. cm.
   Includes index.
   ISBN 978-1-934026-05-2
 1. Mathematics--Study and teaching--United States 2. Mathematics--Study and teaching--
Singapore. 3. Education--Parents participation—United States. 4. Mathematics teachers—In-service
training--United States. I. Title.

  QA13.C444 2008
  510.71—dc22

                2008003250

Editor: Sharon Smith
Art Director and Designer: Soosen Dunholter
Production Coordinator: Deborah Fredericks
Illustration: Marci McAdam
Cover photographs: Nicole Logan

Contents

Acknowledgments

Thank you:

To my late mother, Susan, for always telling me that I could do anything if I just put my mind to it.

To my father, Lenny, for listening to me whine after long days at school.

To my brother and sister, Steven and Lindsay, for being a never-ending comfort.

To my mother-in-law, Judy, for her enthusiastic support of this new adventure.

To Mike Pfister for providing me with the opportunity to work with Staff Development for Educators.

To Lorraine Walker, Char Forsten, Amy Aiello, and Sharon Smith for giving me this wonderful opportunity and allowing me to bounce ideas off them.

To my great friends and fellow faculty members in South River, New Jersey.

To all my students, past and present, for inspiring me to inspire them!

Turning Complaints into Cooperation

As I walked into my classroom at 7:15 a.m., I was greeted by a blinking phone light: "Message Waiting." I put my bags down, grabbed a notepad and pen, and went to check my messages. Usually checking messages was a quick and painless process. "Tommy will be absent today." "Sally is going home with Jan." "Please have Ryan pick up his brother's homework." Not this morning.

"Hello, Mrs. Chen. This is Sam's mom. He does not have his math homework today, nor will he have it until you send home something I can understand. What happened to 'traditional' American math? I don't understand this stuff and don't care to learn it. If you want him to learn a different way to solve problems, teach him at school and keep it at school. Have a great day." CLICK.

Amazed at what I'd just heard, I immediately began calling other teachers into my room to listen to the message, hoping that my interpretation of it was incorrect. Unfortunately, it was not. This was only the second week of school, and already parents were giving me a hard time. Great! Worse yet, the messages did not stop there. More parents began calling in, and not just to me. Teachers throughout the school were having a difficult time getting parents on board. Without their support, the program was doomed to fail.

Does this scenario sound familiar? Anyone who's been there knows that implementing the Singapore Math program can be difficult at first. You have plenty to do already: creating lessons, making manipulatives for activities, and helping your students learn new skills. The last thing you need is a bunch of disgruntled parents calling and leaving you nasty messages or sending in equally awful notes attached to homework, work sheets, quizzes, and tests.

The good news is that you can change that. What we found at my school, and what reports from other schools confirm, is that getting parents involved and on board can be as simple as providing them with the appropriate background information and materials. The Singapore Math program is scary to many parents because some of the strategies are so different from the ones they learned when they were in school. A little communication from you can help them to understand how the strategies work and to see for themselves how effective those strategies can be. And a little understanding can mean a lot of support. The bottom line is that hosting parent nights and sending home work sheets with step-by-step instructions not only can help the parents but also can help keep you sane throughout the year.

In the following pages you'll find ideas for three key ways to reach parents: Icebreakers get parents "hooked" during a back-to-school night. Send-homes explain new strategies and supply practice problems, and parent nights introduce concepts in person and give parents a chance to become comfortable with new strategies as they practice those strategies with teacher guidance. These three approaches may appear to be very different entities, but they all work together to give parents a complete view of the program and help it succeed.

Because I know that none of us ever has enough time, a lot of this book consists of reproducible pages ready for you to copy and put to immediate use. The descriptions of icebreakers, send-homes, and parent nights refer to specific reproducibles, but please feel free to mix and match activities and reproducibles or to modify any of the reproducibles to meet your specific needs.

Once you've settled on which pieces you want to use, you need to decide when to use them. The timing of the back-to-school night is easy, since your district probably has that scheduled into the calendar for you. But what about the send-homes? I've found it's helpful to get these in parents' hands at the same time I introduce a new concept in class. That way, students are not coming home with new information before they've even begun working with it in class, driving everyone nuts as parents try to learn it before you've taught it.

The parent nights can be the trickiest components to schedule because they require the support of multiple people in your district. Luckily, there is no set time frame for when you need to host any of these programs. Different grades will be working on different skills at different times, so it's almost impossible to coordinate these events with the curriculum. Your best bet is to offer the Introduction to Singapore Math as close as possible to the beginning of the school year and then spread the rest out, hosting one every month or every other month.

Ideally, I suggest you begin implementing these pieces at the start of an adoption or supplementation, because that's when they'll be most beneficial. If your school is new to Singapore Math, you may want to try and host your first information night *before* the start of the school year and then continue with others throughout the rest of the year. It's also helpful to keep on holding parent nights throughout following years for those who are new to your district or who could use a refresher and skill builder as their children continue through the grades.

If your district has already implemented Singapore Math, I'd still encourage you to take advantage of the strategies and reproducibles in these pages. Keeping parents informed is necessary regardless of when you began using the program, and all of these components can significantly aid in getting parents on your side even after the fact! After all, in the end we're all working toward one goal: student achievement. Research shows that, typically, the Singapore Math program not only raises student achievement but also promotes number sense—something that's key to the enjoyment of learning mathematics. You know that already, though. Now you just need to get parents to understand. This book is intended to help you get there.

A Quick Note

We've been using the Singapore Math program in my district for three years. We still continue to host parent nights throughout the year in order to help parents experience the impact the Singapore strategies can have and see the success we're achieving in school.

POTENTIAL CHALLENGES

It can be a challenge getting parents on board as you implement any new program, and Singapore Math is no exception. Here are a few situations you may run into—and some suggestions for how to deal with them.

Parents show up for an informational parent night without signing up ahead of time, leaving you with too few materials for the group.

For a general information night such as the Introduction to Singapore Math or Model Drawing night, it's good to come prepared with extra copies of your handouts. As extra insurance, have another teacher on hand that night to make extra copies if they're needed. If you have leftovers at the end of the evening, save them to use later as send-homes.

You run a make-and-take parent night and parents show up without signing up.

This one's more difficult; creating all the materials for a make-and-take can be time-consuming, so you probably won't want to create a lot of extras. (It's still a good idea to make a *few* more, just to allow for Murphy's Law.) The best way to address this is to focus on it ahead of time by stressing that parents are required to send in the sign-up sheet if they're going to attend. Then make a list of the parents who *have* signed up. Keep the list at the front entrance and mark off the parents as they arrive. If parents who aren't on the list show up, you can ask them to wait until everyone else has arrived and then see if there are extra materials. Or ask them to follow along with a friend and then get their own materials from you at a later time.

Parents do not understand your send-homes.

A couple of years ago, I had a parent—I'll call her "Mrs. Jones"—who just did not "get" model drawing. I sent home practice problems each time we worked on a new type of problem, and Mrs. Jones would always return the handout with a note saying she didn't understand it. I asked her student to go home and "teach" her, but that just made things worse. When something like this happens, your best bet is to have the parent in for a "personal tutoring session." I know it sounds like a pain, but it can be very helpful. I had Mrs. Jones in after school one day, and we spent about 30 minutes practicing a couple of problems. She left feeling more comfortable and confident with the process, and after that she was more open to trying new types of problems on her own.

Parents don't speak English.

My district currently has a large number of ESL students whose parents aren't fluent English speakers. In order to help them better understand the program, we have some of our send-homes and handouts translated into their primary language, just as we would do with district notices. This not only helps the parents but also helps the students. For parent nights, you might also want to include translators or have bilingual teachers present alongside the English speakers.

Making the Most of Back-to-School Night

Back-to-school night can be a very scary event for everyone involved. Parents may be meeting the teacher for the first time, and some will come with preconceived notions (not always accurate ones!) about the way the classroom runs, the teacher's qualifications, and the current curriculum. On rare occasions, some parents come ready to fight! Sometimes it seems that every parent's first reaction to Singapore Math strategies, which are generally new to them and often not explained very thoroughly in the textbooks, is sheer frustration. That in turn frustrates the students and the teachers. Other times parents are more quiet, but it's easy to see that they're overwhelmed and confused. Either way, back-to-school night is your chance to cut the confusion and the frustration. Most important, it's your chance to hook parents.

At back-to-school night, parents are interested in finding out about the teacher, the workings of the classroom, and the curriculum and materials being used to teach their students. Even if they've come to complain, your best shot at disarming them and winning them over to your (Singapore) side is to give them the information they need to understand the program. But before you can do that, you need to hook them. Having parents participate in an icebreaker activity based on one of the program's strategies is a wonderful way to start. Once parents see how powerful these strategies are, they usually want to learn more and are likely to be more receptive to future send-homes and parent-night invitations. Here are some icebreaker activities to consider for your next back-to-school night.

ICEBREAKER #1

Model Drawing

Time to allow: 10–15 minutes per problem
Skip this one if: your time is limited

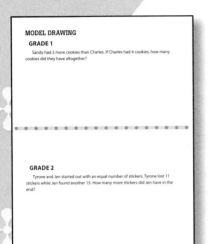

MODEL DRAWING
GRADE 1
Sandy had 2 more cookies than Charles. If Charles had 4 cookies, how many cookies did they have altogether?

GRADE 2
Tyrone and Jen started out with an equal number of stickers. Tyrone lost 11 stickers while Jen found another 15. How many more stickers did Jen have in the end?

If you have enough time, this icebreaker is a great one to use because it gets parents hooked right away with one of the key strategies of Singapore Math—and one that has probably already frustrated some of them.

Prepare ahead of time by making a copy of the reproducible on page 37 showing the steps of model drawing or by writing down the steps your school is using. On a separate sheet, write out the rubric you're using to grade students' drawings. Make enough copies of each so that every parent can have a set or simply post the steps and the rubric at the front of the room. Choose one of the model drawing problems on pages 38–42 or 50–86 and copy it onto an overhead or write it on the board.

On back-to-school night, work through the word problem as if you were working with your students, being sure to question parents and "milk" the problem for all it's worth.

ICEBREAKER #2

Grading a Model-Drawing Solution

Time to allow: 10 minutes or less

STEP-BY-STEP MODEL DRAWING

1. Read the problem.

2. Identify the variables—the "who" and the "what."

3. Draw a unit bar to model each variable.

4. Chunk the problem and adjust your unit bars to match your information. Fill in your question mark.

5. Work your computation.

6. Write a complete sentence to answer the question.

This icebreaker will allow parents to better understand your grading system when tests and quizzes go home and will allow them to help their students complete word problems accurately when at home. (It's also a great activity to try with your students, allowing them to see how you're grading their tests and quizzes.)

Prepare ahead of time by making a copy of the reproducible on page 37 showing the steps of model drawing or by writing down the steps your school is using. On a separate sheet, write out the rubric you're using to grade students' drawings. Make enough copies of each so that every parent can have a set, or simply post the steps and the rubric at the front of the room. Next, choose two of the word problems from pages 38–42 or 50–86 and either copy each one onto a separate overhead transparency or write it on large chart paper or the board. Then write out model drawing solutions to the problems so that as parents arrive, they see the solutions. One of the problems should be solved correctly, allowing for a perfect 10 out of 10 points. In the second problem, leave out one or more components so that the solution will earn a less-than-perfect score.

As the parents arrive, give them the handouts (if any) you're using. Have them help you determine scores for both problems, discussing why each portion does or does not receive points.

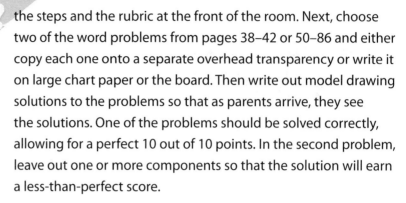

ICEBREAKER #3

COMPUTATION STRATEGY PRACTICE
QUESTIONING SEQUENCE FOR REARRANGING

EXAMPLE: *637 X 5*

1. When multiplying using rearranging, which place value do we start with? *The largest place value—in this case, the hundreds.*
2. What will we be multiplying first? *600 X 5 .*
3. What is 600 X 5? *3,000.*
4. Keep 3,000 in your head and move to the next place value, the tens.
5. What will you be multiplying? *30 X 5.*
6. What is 30 X 5? *150.*
7. What number are you holding in your head? *3,000.*
8. What is 3,000 + 150? *3,150.*
9. Keep 3,150 in your head and move to the next place value, the ones.
10. What will you be multiplying? *7 X 5.*
11. What is 7 X 5? *35.*
12. What number are you holding in your head? *3,150.*
13. What is 3,150 + 35? *3,185.*

Mental Math

Time to allow: no more than 5 minutes per problem, including your explanation of the strategy

As you know, mental math is an important part of Singapore Math. It's also something that can be introduced quickly and can be a real eye-opener for parents.

To prepare, choose one of the computation questioning sequences from pages 117–18. On the board, write a problem that's appropriate for demonstrating that strategy.

On back-to-school night, discuss with parents how you would solve a problem based on the strategy. Run them through a practice problem, being sure to keep your questioning quick in order to reinforce math-fact fluency. Also make sure to explain why that particular strategy is an important one.

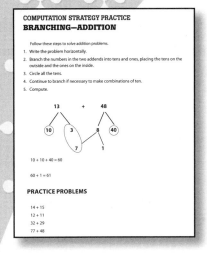

ICEBREAKER #4

Branching

Time to allow: about 5–10 minutes, including the explanation

Branching is another Singapore Math strategy that's new to many parents, and we all know how frustrating it can be to try to teach something you don't understand yourself. You can defuse the tensions associated with that situation by walking parents through the process.

To prepare for this icebreaker, make handouts from the reproducible on page 107 and create an overhead of that same page. Decide on the problem you want to use as your example; it should be one that calls for adding together two two-digit addends.

Give the parents their handouts as soon as they arrive, and then write a problem on the board. Explain the process of branching, being sure to emphasize its importance in reinforcing place value as well as its role in helping students solve addition problems with larger numbers. Following the steps in the handout, use branching to solve the problem with the parents. Then have parents work on another problem on their own. Make sure to review the solution with the group to assess understanding.

ICEBREAKER #5

Area Model for Multiplication

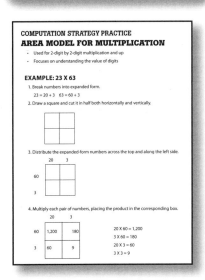

Time to allow: about 5 minutes per problem, including the explanation

This is a good icebreaker for parents of older students because it's quick, it deals with material more common to the older grades, and it shows off something besides the model drawing that parents sometimes think is all there is to Singapore Math.

Prepare for this one by creating handouts from the reproducibles on pages 111–12.

On back-to-school night, distribute the handouts as parents arrive. Then put a two-digit by two-digit multiplication problem on the board. Explain the importance of understanding the value of digits and understanding what multiplication is. Walk through the area model for multiplication, following the steps in the handout and explaining each step as you go along. Then give parents the opportunity to solve a problem on their own. As always, make sure you review the solution with the parents.

Place Value Strips

Time to allow: about 5 minutes
Skip this one if: you don't have enough place value strips for everyone (or at least for every family)

This activity can be a fun one to try with parents because it gets them moving around and interacting with each other. The only preparation needed is to pull together your sets of place value strips.

On back-to-school night, give each participant a set of place value strips. (These are available from Crystal Springs Books— www.crystalsprings.com—or you can use homemade versions based on the reproducibles on pages 96–102.) Ask each parent to create a three- or four-digit number using his strips. Have everyone move around the room, placing themselves in a row in increasing order. Then pull two parents out of the row and ask the rest to compare the numbers those two parents are holding—determining which number is greater than the other and which one is less than the other. Try to get someone to volunteer an explanation that includes the value of the digits. Make sure to note that this is an activity that you do with the students as reinforcement for place value, value of digits, and comparing and ordering numbers.

PLACE VALUE STRIPS

1,000
2,000
3,000
4,000

10 40
20 50
30 60

Sample Send-Homes

When you provide families with constant information regarding the topics you're covering and the strategies you're using, you help parents to help their children. You also create a bond between teacher, parent, and student. A great way to accomplish all of this is to send home handouts whenever you introduce a new topic, giving parents a step-by-step breakdown of what you're doing in class along with practice problems for them to try. Giving them this information not only keeps them informed, but also gives them the tools to reinforce the strategies you're teaching in the classroom.

As the teacher, you get to choose the topics you would like to create send-homes for. It's important to recognize that some of the strategies being taught in the Singapore Math program are not familiar to many parents; these strategies should definitely be included as send-homes. However, some of the more traditional concepts can also be made into send-homes. The purpose of these handouts is to provide parents with clear and timely information to help them work with their children more effectively. You get to decide the best way to meet that goal.

When choosing what you will send home, try to spread out the information. You don't want to send home information on four different topics at one time; that would probably get you nothing except overwhelmed parents. Your best bet is to try and spread out the handouts to coincide with your instruction. (If your textbook includes back-to-back topics that warrant send-homes, try to deal with more commonly understood skills before newer skills that are associated specifically with Singapore Math.) That way you can introduce parents to important concepts at the same time their students are learning those concepts.

Each of the following ideas for send-homes corresponds to one or more specific reproducibles in this book. To fine-tune things for the grade you teach, you may want to modify the reproducibles referenced here, create your own send-homes from scratch, or use some of the other work sheets in the back of this book. All I can say is: once you figure out what will work best for you, go for it! I've included answer keys for reference, but you probably will *not* want to send *those* home. Let the parents and students practice on their own!

SEND-HOME #1

Model Drawing

When you introduce model drawing, send home a copy of the reproducible on page 37 (or substitute a list of the steps you're using for this strategy). From then on, each time you teach your students how to use model drawing with a new type of problem, it's important to send a completed example home to the parents. Give one problem at the top of the page, along with its solution. Then add a second, very similar problem with space for the parents to solve it. Send this work sheet home for parents to complete either on their own or with their students. Ask parents to send the sheet back only if they would like feedback as to whether or not their model was correct. These model drawing send-homes are key to helping parents help students. Review a few of the model drawing send-homes on pages 44–47. Use these to start with and then create some on your own!

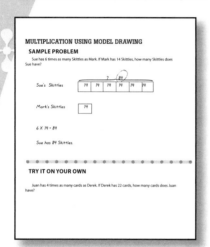

SEND-HOME #2

Branching

Branching is a technique that may seem foreign to many parents. If you send students home with problems and request they solve the problems using branching, you may confuse parents who haven't been introduced to this strategy. Before giving homework assignments requiring the use of branching, send home a work sheet breaking down the steps for branching and giving an example. If you like, use the reproducible on page 107 to explain the use of branching for addition. You might want to follow up later with branching for subtraction (page 109).

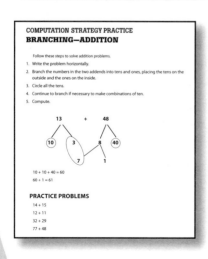

The Parent Connection for Singapore Math

Area Model for Multiplication

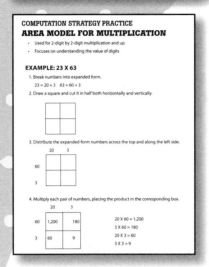

The area model is an effective strategy for teaching large-number multiplication and for reinforcing place value and number sense. Although it may appear to be a simple strategy, it's not familiar to most adults, and so it should be introduced to parents. You might want to begin with the reproducibles on pages 111–12 and then follow up later with further examples using your own numbers.

Visual Model for Adding Fractions with Unlike Denominators

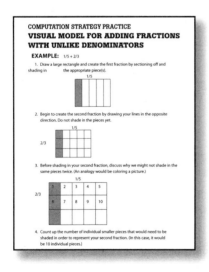

Adding fractions with unlike denominators can be challenging for students who are unable to find common denominators. The visual model strategy can help them to understand the concept in more concrete terms, but, as you know, this strategy can seem difficult at first and requires a careful explanation. It may also be difficult for students to explain this strategy to their parents. Introducing the visual model to parents through a send-home is the ideal solution. You might want to begin with the reproducibles on pages 114–15 and then send home further examples as you see fit.

Parent Nights that Work

Parent nights are among the best tools you have for getting parents on board with your Singapore Math program. They give parents a chance to raise all their questions and voice their frustrations, and they give *you* a chance to show off the strategies and respond to everyone's concerns. Parents won't buy into something they don't understand. This is your chance to win their understanding and their support. Always remember that that's your primary goal.

Organizing a parent night can be fun and stressful at the same time. Depending on the size of your school or district, you may choose to make your parent nights be grade specific or group all your school's parents together. If you're working with a large group, you may want to have multiple teachers available to help out with multiple grade levels. In my school, we've found that by having one key presenter and then also including other "helpers" around the room, we reduce the number of times the main presenter has to stop and explain a skill to one specific person.

In addition to ideas for a parent night that introduces families to Singapore Math, this chapter outlines events that focus on individual aspects of the program: model drawing, place value, and specific computation strategies. I've also included suggestions for make-and-take nights, because those help everyone to see what a huge part manipulatives play in the Singapore Math program.

That doesn't mean you have to offer every one of these. If you can't hold all of the parent nights in this chapter, you might want to begin with the Introduction to Singapore Math Night and the Model Drawing Night, and then schedule whichever others seem most helpful. Pick the ones that look good to you or combine a couple. You can run the Understanding Place Value session and the Computation Strategies session together, for example, and the two related make-and-takes are also logical ones to combine.

Note, too, that the programs don't depend on each other. There's no law that says they have to be presented in the order they appear here (although if you're going to present the make-and-takes, it helps to hold those before Understanding Place Value) or that parents have to attend every session. Make sure everyone knows that if they miss one parent night, they're still welcome at the next one.

I *would* encourage you to include at least one make-and-take if you can. Parents need to know that their students are engaging in hands-on activities on a daily basis. With make-and-takes, you get a chance to point out that each math period has a scheduled activity time in which students work with a vast array of tools and that these tools help them gain a greater understanding of numbers and the specific concepts they're learning. Better yet, make-and-takes give parents a chance to try out the manipulatives for themselves and see exactly how they're used. And the events can be fun!

The idea of a make-and-take is for parents and students to work together to create manipulatives a lot like the ones you're using in class. Including the students in the make-and-takes is great because they can demonstrate to their parents the skills they have learned through Singapore Math. Once families have created the manipulatives together, students will be able to use them at home to complete assignments and study for tests—and parents will be able to help because they'll know how the manipulatives are to be used.

There are tons of manipulatives families can create at make-and-take sessions so that students have them at home to use with their homework. Place value cubes, fractions rods, and number cards are just a few possibilities. I've focused here on making place value disks and strips because those are manipulatives we use all the time in class, and I like to be sure parents are familiar with them. Besides, it's great for the students to have their own sets of the disks and strips to work with at home.

A Quick Note

Maybe you don't have the time and support to implement all of these ideas. If you have to choose, go with the send-homes; they give you the quickest way to grab parents.

The only drawback to make-and-takes is that they take a lot of time—both prep time and running time. They involve a lot of materials that you need to prepare, although the reproducibles on pages 88–102 are intended to give you a head start on that. And since they involve a lot of people making a lot of product, the events themselves can take a while. Most of the other parent nights usually run an hour to an hour and a half. With a make-and-take, parents can leave as soon as they've finished their projects, or they can take the rest of their materials home and finish up there, but the whole thing can last an hour and a half to two hours. Plan accordingly!

Once you've decided which parent nights you want to host, you need to be sure parents have plenty of notice of what's coming. I like to send home a notice about a month before each event, to give parents time to arrange their schedules. (You may want to include dates and times for other upcoming events in your note as well.) You can use or modify the reproducibles in the back of the book for this or you can create your own. I usually section off a part of the notice as a response slip to be returned to the school, so I have a sense of how many are coming and—especially for the make-and-takes—can make sure I have enough handouts ready. Because I typically present these programs to groups that span several grade levels, I need to get a sense of how to focus the presentations. That's why the response slip also asks for the name of the student whose parents are coming, the grade the child is in, and the name of the child's teacher. I follow up about a week before each event by sending home a simple reminder notice.

When the response slips come back, if you find that the majority of those who will be coming are parents of students in the lower grades, I recommend

A Quick Note

Parents really get into model drawing. I had one parent yell out in the middle of the event, "I wish I'd learned this when I was in school. Maybe I wouldn't have been so bad at math!" Everyone started laughing, joking around, and having a great time with it!

that you still follow the same outline for each evening even though you may decide to change which pieces you emphasize. Many of the events cover a wide range of materials to help parents with students in all grade levels. Besides, the parents of children in lower grades will be happy to see the more advanced materials and know where their children will be headed in the future.

As you get closer to each parent night, pull together your overheads as well as a packet of handouts for each parent. It helps to include extra sheets for note taking at the back of each packet, and to make sure you have a few extra packets for parents who didn't sign up. If you're planning to hold other parent nights on different topics later, include a notice giving dates and times for those events in the back of each packet, too. A little advance advertising never hurt! For the handouts and other materials needed for each individual session, refer to the specifics in the following pages.

A lot goes into preparing a good parent night, but the events themselves can be fun. They can also be chaotic. Make sure to keep the focus on your key points for each evening, but allow plenty of time for questions at the end or be prepared to stay late to answer questions. (You may want to allow for staying late anyway. If one parent just doesn't "get it" and starts to monopolize the discussion, you can always invite that person to stay and talk with you individually afterward.) If you're not sure of an answer, get the parent's contact information so that you can get back to that person with the appropriate response.

Remember: your objective is to get the parents on your side, and the way to do that is to be sure they understand the strategies. But if you can get them to relax and have fun with it, too, you'll really win them over!

MATERIALS

- Reproducibles on pages 34–37 and your choice from pages 38–42 (minus the answer keys)
- Your rubric for scoring model drawing solutions
- Plain white copy paper
- Blank transparencies
- Overhead markers in multiple colors
- Overhead projector
- Sample place value disks
- Sample place value mats
- Sample place value strips
- Sample place value cubes
- Copies of textbooks and workbooks for the appropriate grade levels

INTRODUCTION TO SINGAPORE MATH

If you hold only one parent night, make it this one. It's vital to give parents background information explaining why your school has chosen to use the Singapore Math program, and you can use this time to do just that. Work with the reproducibles listed, or substitute your own versions, to create handouts and overheads ahead of time. Include just a couple of word problems. (A word to the wise: when you create the overheads for the word problems, it's good to position each reproducible on your copier so you get only one problem per page.) Create handouts and an overhead of your scoring rubric, too.

At the event, explain what the TIMSS report is, present the findings from that report, and explain why those findings provide a valid reason for using Singapore Math strategies in your school. Offer a few key points about what the program involves and what its strengths are, and outline the structure of a typical Singapore Math class.

Beyond that, of course the one thing you *have* to include is model drawing! Explain the steps involved as well as your scoring system and walk the parents through the solutions to the sample problems. This is your bait, and you can't catch a fish without bait!

As the final part of the evening, demonstrate some of the manipulatives you're using in class, particularly the place value disks, mats, strips, and cubes. Giving parents examples of how these manipulatives work can be very helpful. Besides, with any luck, your demonstration will make them want to come to one of the make-and-take events!

The Introduction to Singapore Math can be the trickiest of all the parent nights because it covers such a wide range of material, and you don't want to leave out any important information. You want to sell the program and introduce at least two of its key components: model drawing and understanding of place value. Ideally, you want to be able to answer questions about the textbooks and workbooks, too. That's a lot to cover. But this introductory night is also a great opportunity. Parents may come in the door ready to fight, but they'll leave with a newfound respect for what you do in the classroom.

A Quick Note

At the first Singapore Math parent night we held at my school, parents walked through the door ready to fire question after question. They hit us with some tough ones, but we were always able to supply them with consistent responses. Once we began doing some model drawing and they saw the usefulness of the strategy, they began getting into it. By the end of the night, they were asking if there would be more sessions available to help them better understand the strategies.

PARENT NIGHT #2

MODEL DRAWING

MATERIALS

- *Reproducibles on pages 37 and 50–86 (minus the answer keys)*
- *Plain white copy paper*
- *Blank transparencies*
- *Overhead markers in multiple colors*
- *Overhead projector*
- *Copies of the textbooks, workbooks, and other resources you're using in class*

Model drawing is one of the key strategies embedded in the Singapore Math program, and it's one of the most controversial ones. When choosing which parent nights to run, this should be at the top of your list, second only to the general introduction. This parent night is imperative to the success of the program.

The preparation for this one can be lengthy, because in addition to making overheads and handouts of the reproducibles, and pulling together the other Singapore Math resources in case parents have questions about them, you need to make sure that you're able to solve all of the problems you plan to present to your audience. Prepare yourself for this just as you'd prepare for your students.

Don't just review the answer key. It's important to solve the problems yourself and then check back—and if there's more than one way to solve a particular problem, make sure you can demonstrate and explain each approach. Be prepared to justify the way you're interpreting the information in each problem.

If that sounds a little overwhelming, relax. Have fun with it. Include some sample problems that refer in a light-hearted way to individuals in the group (choosing people you know can handle this, of course). Ask the parents to solve some problems on their own. Call parent volunteers up in front of the crowd to try out their new skills. Above all, don't let the parents get to you. Instead, let the material get to them!

PLACE VALUE DISKS & MATS MAKE-&-TAKE

MATERIALS

- *Reproducibles on pages 88–95*
- *Colored paper or card stock*
- *Plain white copy paper*
- *Sandwich-size resealable plastic bags for carrying manipulatives home*
- *Crayons*
- *Scissors*
- *12 X 18-inch sheets of white construction paper*
- *Glue sticks*
- *Rulers*
- *Laminating machine (if available)*

Place value disks and mats are great manipulatives for students—especially those in the younger grades—to have at home. This make-and-take gives parents hands-on experience with creating those manipulatives and seeing how they're used. Depending on how many parents are involved, though, just copying all the reproducibles for this can be time-consuming, so be sure to allow plenty of preparation time.

To get ready for this event, copy each of the reproducible disk pages onto colored copy paper or card stock for participants so that the disks match the colors of the manipulatives you're using in class. (If you'd rather, you can copy everything onto white paper or card stock and have parents and students color them with crayons in the appropriate colors on the night of the make-and-take.) Then copy the labels for the mats onto white paper. Ideally, make a few extras of everything for last-minute participants. Collate the sets of reproducibles so you can give each family a complete set, along with a plastic bag for carrying purposes, as they arrive at your make-and-take session. Create a set of disks and a mat for parents to use as a model. Have all the other materials available for participants to use.

Before parents arrive, write on the board the instructions for creating the disks and mats. Then list a few practice problems that parents can try solving with the manipulatives.

Once participants have arrived, begin the make-and-take by having parents and students cut out their place value disks. When the disks are ready, have each family create a place value mat, following your model and the instructions below. If possible, laminate the mats for the parents and students to take home.

Finally, give parents a choice: they can leave as soon as they finish creating their materials, or they can use their new disks and mats to practice solving the problems on the board.

DIRECTIONS FOR MAKING PLACE VALUE DISKS & MATS

1. If the disks are on plain white paper, begin by coloring them with the appropriate crayons. (See the sample provided.) Otherwise, begin with Step 2.

2. Cut out the individual place value disks. To make this easier, you may want to just cut them into squares rather than circles.

3. Cut out the headings for the place value mats.

4. Following the sample, place each heading in the appropriate location on a large sheet of construction paper.

5. Glue the headings in place.

6. Draw vertical lines to separate the columns.

Millions	Hundred Thousands	Ten Thousands	Thousands	Hundreds	Tens	Ones
			1,000	100	10	1
			1,000	100	10	1
				100		1
				100		

Time to allow:
1 1/2 to 2 hours

MATERIALS

- *Reproducibles on pages 96–102*
- *White card stock*
- *Crayons or colored pencils*
- *Scissors*
- *Resealable plastic bags for carrying manipulatives home*

PLACE VALUE STRIPS MAKE-&-TAKE

As the name indicates, these strips are key to building an understanding of place value. But it's good for parents to see that students can also use them to practice addition, subtraction, multiplication, division, and comparing and ordering numbers, among other skills. Providing this resource to parents and students is key to continued success at home.

To prepare for this session, copy the reproducibles for the strips onto the card stock and collate the sheets into easily manageable packets. (If you can't print them on card stock, use paper; it's less sturdy but fine.) Color in and cut out a set of strips for parents to use as a model. On the board, list the colors you're using in class for each value.

On the evening of the event, hand out the packets as participants arrive. Provide parents and students with crayons or colored pencils and allow them to color in the pieces appropriately. Then hand out the scissors and plastic bags and have each family cut out their set of place value strips to take home.

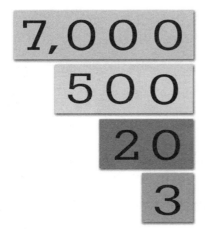

After they create strips like those at left, parents and students combine them as shown below.

Once the strips are completed, give participants a chance to try using them. One option is to have each person create a number with as many digits as you name. Be sure to explain that if a participant decides on a number that has a zero for a particular place value, he must still include that zero when he creates his number. (If the number is 1,305, he must include the 00 card for the tens place value.) Once all the numbers are ready, ask participants to put themselves in increasing order. Pull out individuals to compare their numbers.

For example, let's say each family unit has created a set of strips. Johnny Jones uses some of his family's strips to create the number 1,305, his dad makes 142, and his mom makes 967. Suzy Smith creates 43 and each of her parents comes up with a different number. Everyone in the room lines up in increasing order. You pull Suzy and Mr. Jones and ask the "class" to tell you which number is larger and which is smaller.

Another option for practicing with the strips involves having participants pair off. Asking participants to use their ones place value cards just as if they were a deck of playing cards, have each person randomly pull cards and create a two- or three-digit number. Next, have each person pull a single additional card to create a multiplication problem and then solve that problem. For each pair of players, the person with the larger solution wins a point.

For example, let's say I pull 325 and 7. My multiplication problem is 325 X 7 = 2,275. If my partner has 201 X 3 = 603, I win the point because my number is larger. Repeat this process five times. At the end, whoever has the most points is the winner! You can do this with addition, subtraction, and division as well.

A Quick Note

In class, students have a blast moving around the room with their numbers. Once they've placed themselves in order, we use mental math skills to find the differences between specific numbers (the range). This activity allows us to work on place value, comparing and ordering numbers, and mental math. Even better, students think they're playing a game!

Time to allow:
1 hour (or more if
you include more
examples)

MATERIALS

- *Reproducibles on pages 103–05*
- *Plain white copy paper*
- *Blank transparencies*
- *White card stock*
- *Overhead place value disks, if available*
- *Overhead markers*
- *Overhead projector*
- *Place value disks*
- *Place value strips*

UNDERSTANDING PLACE VALUE

You know that Singapore Math is much more than just model drawing, but parents don't always "get" that. So you need to show them! There are so many other key components of the program that help our students gain a more in-depth understanding of numbers. Showing parents some of the fun ways Singapore Math can be used to reinforce number sense is another great way to hook them. Keep them wanting more by continually showing them something new!

To prepare for this parent night, make overheads of the reproducibles for the place value problems and the place value mat as well as paper copies for each participant. If possible, you may want to enlarge the copies of the mat. Make extra copies of the mat on card stock or paper for parents to take home, too. If you held either of the place value make-and-takes, make extra copies of the reproducibles (pages 88–102) from those sessions now, so that parents who weren't able to make it to those events have the option of taking the copies home and putting them together later.

On the night of the event, arrange tables so parents and students can work at them. Leave sets of manipulatives on each table so anyone who arrives early can "play" with them.

When you're ready to begin, distribute the handouts. One at a time, reveal a problem on an overhead transparency and then use the markers (or the overhead disks) and the transparency of the mat to demonstrate solving that problem. Ask parents to follow along with their own disks and mats. Don't be afraid to have fun with this! Once you've demonstrated the process, ask for a volunteer or two to lead the group by working a problem on the overhead.

To demonstrate the place value strips, you may want to try out any of the activities described on page 14 or 28. Or try some rounding activities. Have parents create numbers from the strips and then round those numbers. Or they can use their strips to create two different numbers and estimate the sum of those two numbers by rounding.

A Quick Note

One way to make sure you have enough time is to start promptly. Those who are on time will appreciate it, and those who arrive late will catch up!

Time to allow:
1 to 1 1/4 hours

COMPUTATION STRATEGIES

MATERIALS

- *Reproducibles on pages 106–18*
- *Plain white copy paper*
- *Blank transparencies*
- *Overhead markers*
- *Overhead projector*

Teaching our students many different ways to solve problems enhances their ability to choose a strategy that works best for them. It's essential to explain to parents that, as part of the Singapore Math program, you will be teaching students some strategies that are new along with many others that they (the parents) are familiar with.

As you prepare for this parent night, it's important to recognize that parents will be asking you to justify each strategy you show them. Be prepared to explain the significance of teaching multiple strategies and to reinforce the importance of practicing these strategies at home along with those that the parents may already know. Also, make sure you're fluent in each of the strategies you're demonstrating, even if it's not one that's taught at your grade level. And of course, take the time to make transparencies of the reproducibles and to create handouts from them, too—including the answer keys if you plan to send those home with participants.

On the night of the event, begin with an easy concept, number bonds, and use the reproducible (page 106) to show the relationship between number bonds and math facts. Point out how an understanding of number bonds helps to build strong number sense.

From there, move on to branching. Working from the reproducible on page 107, put a transparency on the projector and walk the group through an addition problem. Point out that you're pulling the number apart into expanded form and then bringing it back together into standard form in the end; emphasize the fact that you're using addition as the final step. After you demonstrate, let the parents try the practice problems. Repeat the process to teach branching for subtraction (page 109).

Next up is the area model for multiplication. Again walk the parents through the steps on the reproducible (pages 111–12) and then let them practice on their own.

Now present the visual model for adding fractions with unlike denominators (pages 114–15). Follow the same process as for the earlier strategies, and then ask parents to solve these problems:

$$2/3 + 1/4 \qquad 1/7 + 3/5 \qquad 1/5 + 5/6$$

(Hint to give parents: the answer to the last of these problems is an improper fraction/mixed number.)

Once again, you don't need to include these practice problems on your handouts, but you can find a reproducible answer key on page 116 in case you want to send it home.

You now have just two more strategies to "wow" the parents with. One is left-to-right computation (page 117); the other is rearranging (page 118). For each of these, walk the parents through the steps on the appropriate reproducible, making sure you ask the same questions you would ask your students. In the case of left-to-right computation, it's important to mention that it does take time and practice to master this strategy, but that with continuous repetition, the manipulation of the numbers without paper and pencil becomes much easier and faster.

As always, be sure to leave time for questions.

A Quick Note

In about the middle of October, my class was working on a mental-math strategy using three-digit numbers. They became so fluent at it that one of my students asked if she could be the teacher. Of course I said yes, excited to see what would happen. She did a wonderful job "running" mental math. I later decided that every day, I would choose a student to "run" a left-to-right mental math problem regardless of what strategy we were working on that day. They loved it and so did their parents when they found out.

Reproducibles

TRENDS IN INTERNATIONAL MATHEMATICS & SCIENCE STUDY
TIMSS REPORT 2003

4TH GRADE

Singapore	594
Hong Kong	575
Japan	565
Chinese Taipei	564
Belgium	551
Netherlands	540
Latvia	536
Lithuania	534
Russian Federation	532
England	531
Hungary	529
United States	**518**
Cyprus	510
Moldova	504
Italy	503
Australia	499
New Zealand	493
Scotland	490
Slovenia	479
Armenia	456
Norway	451
Iran	389
Philippines	358
Morocco	347
Tunisia	339
International Average	495

SINGAPORE MATH
KEY REFERENCE POINTS

- Language-based math program helping children make connections between pictures, words, and numbers

- Accelerated program, driving all ability levels

- Cumulative program that revisits concepts covered earlier by connecting strands of mathematics

- Topic intensive, with fewer topics covered per grade level

- Smaller textbooks, with skills not re-taught formally

- Mental-math strategies embedded in the program

- Highly visual program that benefits special-needs students and inclusion students

SINGAPORE MATH
CLASS STRUCTURE

- Average period between 60 and 90 minutes
- Contains 5 distinct segments
    - ◊ Mental math   10 min.
    - ◊ Teacher-directed lesson   20 min.
    - ◊ Activity   20 min.
    - ◊ Problem-solving   15 min.
    - ◊ Independent practice   15 min.

STEP-BY-STEP MODEL DRAWING

1. Read the problem.

2. Identify the variables—the "who" and the "what."

3. Draw a unit bar to model each variable.

4. Chunk the problem and adjust your unit bars to match your information. Fill in your question mark.

5. Work your computation.

6. Write a complete sentence to answer the question.

MODEL DRAWING
GRADE 1

Sandy had 2 more cookies than Charles. If Charles had 4 cookies, how many cookies did they have altogether?

GRADE 2

Tyrone and Jen started out with an equal number of stickers. Tyrone lost 11 stickers while Jen found another 15. How many more stickers did Jen have in the end?

ANSWER KEY
MODEL DRAWING

GRADE 1

Sandy had 2 more cookies than Charles. If Charles had 4 cookies, how many cookies did they have altogether?

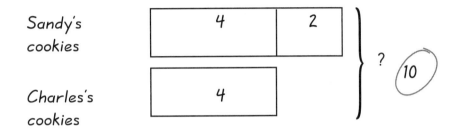

A. 4 + 2 = 6

B. 6 + 4 = 10

Sandy and Charles had 10 cookies altogether.

GRADE 2

Tyrone and Jen started out with an equal number of stickers. Tyrone lost 11 stickers while Jen found another 15. How many more stickers did Jen have in the end?

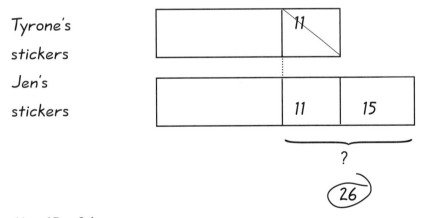

11 + 15 = 26

Jen had 26 more stickers than Tyrone in the end.

MODEL DRAWING
GRADE 3

Lindsay had 3 times as much money as Danny. If they had $40 altogether, how much money did Lindsay have?

GRADE 4

Five-sevenths of the class has more than 1 sibling. If 12 students have only 1 sibling each, how many students are in the whole class?

ANSWER KEY
MODEL DRAWING

GRADE 3

Lindsay had 3 times as much money as Danny. If they had $40 altogether, how much money did Lindsay have?

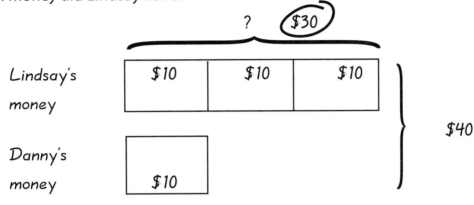

A. 4 units = $40 $40 ÷ 4 = $10   1 unit = $10

B. 3 X $10 = $30

Lindsay had $30.

GRADE 4

Five-sevenths of the class has more than 1 sibling. If 12 students have only 1 sibling each, how many students are in the whole class?

A. 2 units = 12   12 ÷ 2 = 6   1 unit = 6

B. 7 X 6 = 42

There are 42 students in the class.

MODEL DRAWING
GRADE 5

The ratio of cows to pigs is 7 : 3. If there are 32 more cows than pigs, how many pigs are there?

GRADE 6

At the county fair, 40% of the people ate funnel cake, 20% of the people ate gyros, and the rest had snow cones. If 15,260 people attended the county fair, how many people ate funnel cake and gyros altogether?

ANSWER KEY
MODEL DRAWING

GRADE 5

The ratio of cows to pigs is 7 : 3. If there are 32 more cows than pigs, how many pigs are there?

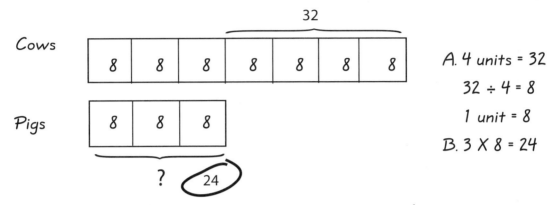

A. 4 units = 32

   32 ÷ 4 = 8

   1 unit = 8

B. 3 X 8 = 24

There are 24 pigs.

GRADE 6

At the county fair, 40% of the people ate funnel cake, 20% of the people ate gyros, and the rest had snow cones. If 15,260 people attended the county fair, how many people ate funnel cake and gyros altogether?

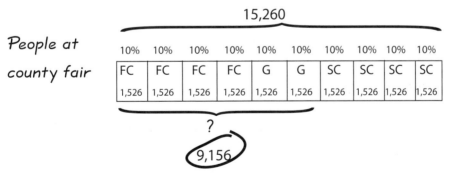

A. 10 units = 15,260   B. 6 X 1,526 = 9,156

   15,260 ÷ 10 = 1,526

   1 unit = 1,526

At the county fair, 9,156 people ate funnel cake and gyros altogether.

ADDITION USING MODEL DRAWING
SAMPLE PROBLEM

On Tuesday, the bakers made 40 loaves of bread. On Wednesday, they made 15 more loaves of bread than they made on Tuesday. How many loaves of bread did they make on Tuesday and Wednesday altogether?

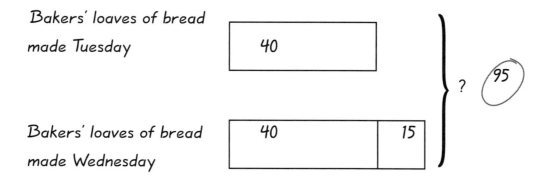

A. 40 + 40 = 80   A. 40 + 15 = 55

OR

B. 80 + 15 = 95   B. 55 + 40 = 95

The bakers made 95 loaves of bread on Tuesday and Wednesday altogether.

● ● ● ● ● ● ● ● ● ● ● ● ● ● ● ● ● ● ● ●

TRY IT ON YOUR OWN

Ling has 48 flowers. Cindy has 25 flowers more than Ling. How many flowers do Ling and Cindy have altogether?

SUBTRACTION USING MODEL DRAWING

SAMPLE PROBLEM

Tom has $65. Jon has $20 less than Tom. How much money does Jon have?

Tom's money

$65	

Jon's money

?	$45	$20

$65

$65 - $20 = $45

Jon has $45.

● ● ● ● ● ● ● ● ● ● ● ● ● ● ● ● ● ● ● ●

TRY IT ON YOUR OWN

Sally has 64 crayons. Janey has 40 fewer crayons than Sally. How many crayons does Janey have?

MULTIPLICATION USING MODEL DRAWING

SAMPLE PROBLEM

Sue has 6 times as many Skittles as Mark. If Mark has 14 Skittles, how many Skittles does Sue have?

Sue's Skittles

| 14 | 14 | 14 | 14 | 14 | 14 |

Mark's Skittles

| 14 |

6 X 14 = 84

Sue has 84 Skittles.

TRY IT ON YOUR OWN

Juan has 4 times as many cards as Derek. If Derek has 22 cards, how many cards does Juan have?

FRACTIONS USING MODEL DRAWING

SAMPLE PROBLEM

Two-fifths of the children have green backpacks. If 12 children have green backpacks, how many children are there altogether?

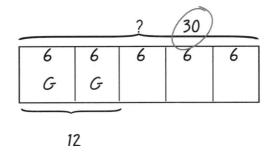

A. 2 units = 12

12 ÷ 2 = 6

1 unit = 6

B. 5 X 6 = 30

There are 30 children altogether.

TRY IT ON YOUR OWN

At the ice skating rink, 5/8 of the skaters were children. If there were 45 children at the skating rink, how many people were at the rink altogether?

ANSWER KEY
MODEL DRAWING

ADDITION

Ling has 48 flowers. Cindy has 25 flowers more than Ling. How many flowers do Ling and Cindy have altogether?

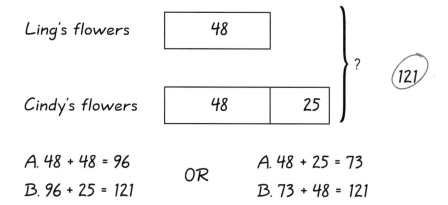

A. 48 + 48 = 96   A. 48 + 25 = 73

OR

B. 96 + 25 = 121   B. 73 + 48 = 121

Ling and Cindy have 121 flowers altogether.

SUBTRACTION

Sally has 64 crayons. Janey has 40 fewer crayons than Sally. How many crayons does Janey have?

Sally's crayons | 64

Janey's crayons | ? (24) 40
64

64 – 40 = 24

Janey has 24 crayons.

ANSWER KEY
MODEL DRAWING

MULTIPLICATION

Juan has 4 times as many cards as Derek. If Derek has 22 cards, how many cards does Juan have?

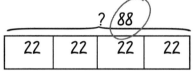

Juan's cards

Derek's cards

4 X 22 = 88

Juan has 88 cards.

FRACTIONS

At the ice skating rink, 5/8 of the skaters were children. If there were 45 children at the skating rink, how many people were at the rink altogether?

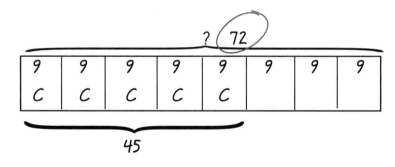

A. 5 units = 45

45 ÷ 5 = 9

1 unit = 9

B. 8 X 9 = 72

There were 72 people at the rink altogether.

MODEL DRAWING
ADDITION

Pam has 3 red, 2 purple, and 4 blue flowers in her garden. How many flowers does Pam have in her garden altogether?

ADDITION

Tony has 12 baseball cards. Patrick has 4 more baseball cards than Tony. How many baseball cards do Tony and Patrick have altogether?

ANSWER KEY
MODEL DRAWING

ADDITION

Pam has 3 red, 2 purple, and 4 blue flowers in her garden. How many flowers does Pam have in her garden altogether?

Pam's red flowers

Pam's purple flowers

Pam's blue flowers

? ⑨

3 + 2 + 4 = 9

Pam has 9 flowers in her garden.

ADDITION

Tony has 12 baseball cards. Patrick has 4 more baseball cards than Tony. How many baseball cards do Tony and Patrick have altogether?

Tony's baseball cards | 12

Patrick's baseball cards | 12 | 4

? ㉘

A. 2 X 12 = 24
B. 24 + 4 = 28

Tony and Patrick have 28 baseball cards altogether.

MODEL DRAWING
ADDITION

The bakers sold 250 bagels on Monday morning. They sold 225 bagels on Tuesday morning. How many bagels did the bakers sell altogether on Monday and Tuesday?

ADDITION

Carlos earns a weekly salary of $500. Thomas's weekly salary is $95 more than Carlos's weekly salary. Joan's weekly salary is $50 more than Thomas's weekly salary. How much money do the three earn altogether on a weekly basis?

ANSWER KEY
MODEL DRAWING

ADDITION

The bakers sold 250 bagels on Monday morning. They sold 225 bagels on Tuesday morning. How many bagels did the bakers sell altogether on Monday and Tuesday?

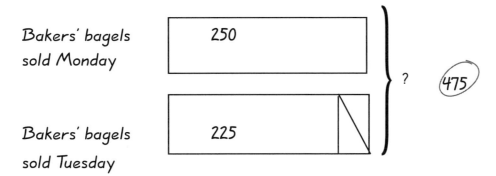

$$250 + 225 = 475$$

The bakers sold 475 bagels altogether on Monday and Tuesday.

ADDITION

Carlos earns a weekly salary of $500. Thomas's weekly salary is $95 more than Carlos's weekly salary. Joan's weekly salary is $50 more than Thomas's weekly salary. How much money do the three earn altogether on a weekly basis?

Carlos's weekly salary — $500

Thomas's weekly salary — $500 $95

Joan's weekly salary — $500 $95 $50

? $1,740

A. $500 + $95 = $595   C. $500 + $595 = $1,095
B. $595 + $50 = $645   D. $1,095 + $645 = $1,740

Carlos, Thomas, and Joan earn a combined weekly salary of $1,740.

MODEL DRAWING
SUBTRACTION

Mia has $100. She buys a dress for $45. How much money does Mia have left?

SUBTRACTION

Mr. Smith earns $3,500 a month. Each month he uses $2,000 to pay his bills and puts $1,000 in the bank. How much money does Mr. Smith have left over each month?

ANSWER KEY
MODEL DRAWING

SUBTRACTION

Mia has $100. She buys a dress for $45. How much money does Mia have left?

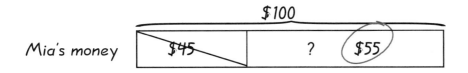

$100 - $45 = $55

Mia has $55 after buying her dress.

SUBTRACTION

Mr. Smith earns $3,500 a month. Each month he uses $2,000 to pay his bills and puts $1,000 in the bank. How much money does Mr. Smith have left over each month?

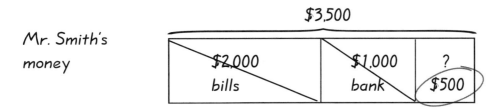

A. $2,000 + $1,000 = $3,000

B. $3,500 - $3,000 = $500

Mr. Smith has $500 left over each month.

MODEL DRAWING
SUBTRACTION

There were 325 video games at the video game store. On Friday, the store sold 32 games. On Saturday, the store sold 54 games. How many games were left in the video game store?

SUBTRACTION

Julio started with a box of 300 crayons on the first day of school. On the last day of school, Julio counted his crayons and found he had only 176 crayons left. How many crayons did Julio lose during the school year?

ANSWER KEY
MODEL DRAWING

SUBTRACTION

There were 325 video games at the video game store. On Friday, the store sold 32 games. On Saturday, the store sold 54 games. How many games were left in the video game store?

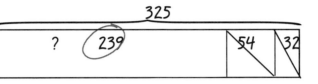

Store's video games

A. 54 + 32 = 86

B. 325 - 86 = 239

There were 239 games left in the video game store.

SUBTRACTION

Julio started with a box of 300 crayons on the first day of school. On the last day of school, Julio counted his crayons and found he had only 176 crayons left. How many crayons did Julio lose during the school year?

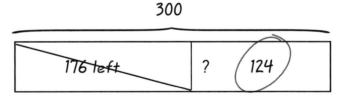

Julio's crayons

300 - 176 = 124

Julio lost 124 crayons during the school year.

MODEL DRAWING
MULTIPLICATION

When planting her garden, Tanya placed her flowers in 5 rows. She put 6 flowers in each row. How many flowers did Tanya plant in her garden?

● ●

MULTIPLICATION

Each student in the class came to school with 8 pencils. If there were 10 students in the class, how many pencils did they have altogether?

ANSWER KEY
MODEL DRAWING

MULTIPLICATION

When planting her garden, Tanya placed her flowers in 5 rows. She put 6 flowers in each row. How many flowers did Tanya plant in her garden?

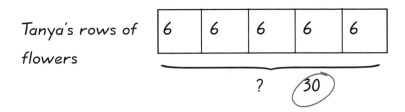

5 X 6 = 30

Tanya planted 30 flowers in her garden.

MULTIPLICATION

Each student in the class came to school with 8 pencils. If there were 10 students in the class, how many pencils did they have altogether?

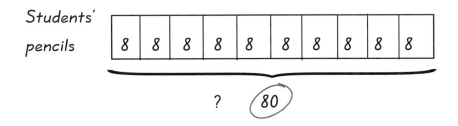

10 X 8 = 80

The students had 80 pencils altogether.

MODEL DRAWING
MULTIPLICATION

Danny made $15 cutting his neighbor's grass each week. If Danny saved his money for 8 weeks, how much money would he save?

MULTIPLICATION

The movie theater has 25 rows of chairs with 36 chairs in each row. How many chairs are in the movie theater?

ANSWER KEY
MODEL DRAWING

MULTIPLICATION

Danny made $15 cutting his neighbor's grass each week. If Danny saved his money for 8 weeks, how much money would he save?

Danny's money saved

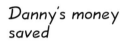

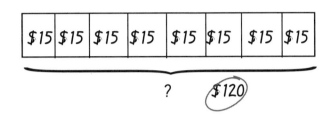

8 X $15 = $120

Danny would save $120 in 8 weeks.

MULTIPLICATION

The movie theater has 25 rows of chairs with 36 chairs in each row. How many chairs are in the movie theater?

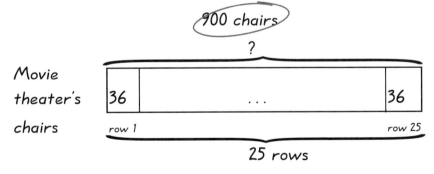

25 X 36 = 900

The movie theater has 900 chairs.

MODEL DRAWING
DIVISION

A deck of 52 cards was divided evenly among 4 players. How many cards did each player receive?

DIVISION

The bagel shop delivered 350 bagels on Monday morning, dividing them among 7 different businesses. If each business received the same number of bagels, how many bagels did each business receive?

ANSWER KEY
MODEL DRAWING

DIVISION

A deck of 52 cards was divided evenly among 4 players. How many cards did each player receive?

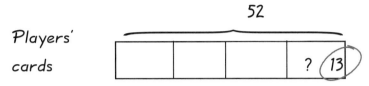

4 units = 52

52 ÷ 4 = 13

1 unit = 13

Each player received 13 cards.

DIVISION

The bagel shop delivered 350 bagels on Monday morning, dividing them among 7 different businesses. If each business received the same number of bagels, how many bagels did each business receive?

350

Bagel
shop's
bagels

? 50

7 units = 350

350 ÷ 7 = 50

1 unit = 50

Each business received 50 bagels.

MODEL DRAWING
DIVISION

Sophia's mom baked 56 cookies for her to bring to school. Sophia and her 7 friends shared the cookies evenly. How many cookies did each of the girls have?

DIVISION

Julio brought 105 Yu-Gi-Oh cards to school to play a game. Five boys, including Julio, played the game. Each boy started with the same number of cards. How many cards did each boy start with?

The Parent Connection for Singapore Math

ANSWER KEY
MODEL DRAWING
DIVISION

Sophia's mom baked 56 cookies for her to bring to school. Sophia and her 7 friends shared the cookies evenly. How many cookies did each of the girls have?

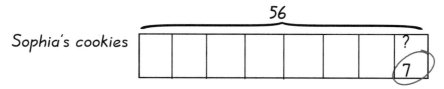

8 units = 56

56 ÷ 8 = 7

1 unit = 7

Each girl had 7 cookies.

DIVISION

Julio brought 105 Yu-Gi-Oh cards to school to play a game. Five boys, including Julio, played the game. Each boy started with the same number of cards. How many cards did each boy start with?

Julio's
Yu-Gi-Oh
cards

105

?
21

5 units = 105

105 ÷ 5 = 21

1 unit = 21

Each boy started with 21 cards.

MODEL DRAWING
FRACTIONS

Four-sevenths of a class of students are girls. If there are 16 girls in the class, how many boys are in the class?

FRACTIONS

There are 36 marbles in a jar. Two-ninths of the marbles are red. There are 8 red marbles. How many marbles are *not* red?

ANSWER KEY
MODEL DRAWING
FRACTIONS

Four-sevenths of a class of students are girls. If there are 16 girls in the class, how many boys are in the class?

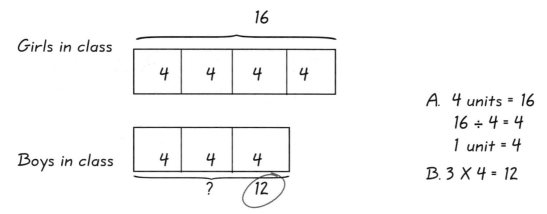

A. 4 units = 16
   16 ÷ 4 = 4
   1 unit = 4
B. 3 X 4 = 12

There are 12 boys in the class.

OR

A. 4 units = 16
   16 ÷ 4 = 4
   1 unit = 4
B. 3 X 4 = 12

There are 12 boys in the class.

ANSWER KEY
MODEL DRAWING
FRACTIONS

There are 36 marbles in a jar. Two-ninths of the marbles are red. There are 8 red marbles. How many marbles are *not* red?

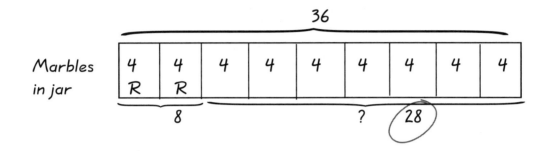

A. 2 units = 8

8 ÷ 2 = 4

1 unit = 4

B. 7 X 4 = 28

Twenty-eight marbles are not red.

MODEL DRAWING
FRACTIONS

Tim saved $600 to buy a new bicycle and helmet. He spent 2/5 of his money on the bicycle. He then spent 1/4 of his remaining money on the helmet. How much did Tim spend on the bicycle and helmet altogether? How much money did he have left over?

DECIMALS

Mrs. Chen bought 4 crayon cases and 8 boxes of crayons for the upcoming school year. The cost of a box of crayons was $.50 more than the cost of a crayon case. Mrs. Chen spent $40 altogether. What was the cost of 1 crayon case? What was the cost of 1 box of crayons?

ANSWER KEY
MODEL DRAWING
FRACTIONS

Tim saved $600 to buy a new bicycle and helmet. He spent 2/5 of his money on the bicycle. He then spent 1/4 of his remaining money on the helmet. How much did Tim spend on the bicycle and helmet altogether? How much money did he have left over?

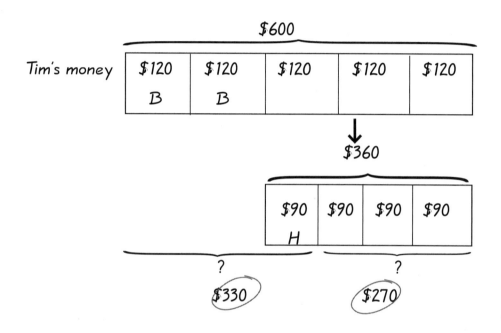

A. 5 units = $600 $600 ÷ 5 = $120   1 unit = $120

B. 3 X $120 = $360

C. 4 units = $360 $360 ÷ 4 = $90   1 unit = $90

D. (2 X $120) + $90 = $330

E. 3 X $90 = $270

Tim spent $330 on the new bicycle and helmet. He had $270 left over.

ANSWER KEY
MODEL DRAWING
DECIMALS

Mrs. Chen bought 4 crayon cases and 8 boxes of crayons for the upcoming school year. The cost of a box of crayons was $.50 more than the cost of a crayon case. Mrs. Chen spent $40 altogether. What was the cost of 1 crayon case? What was the cost of 1 box of crayons?

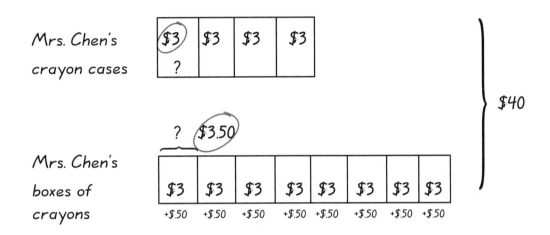

A. 8 X $.50 = $4

B. $40 - $4 = $36

C. 12 units = $36

   $36 ÷ 12 = $3

   1 unit = $3

D. $3 + $.50 = $3.50

The cost of 1 crayon case was $3. The cost of 1 box of crayons was $3.50.

MODEL DRAWING
DECIMALS

Charlene bought 8 bowls and 8 plates for her new apartment. Each plate cost $.60 more than each bowl. Charlene spent $60.80 for the bowls and plates altogether. What was the cost of 1 bowl? What was the cost of 1 plate?

RATE

A car traveled 390 miles in 6 hours. If the car drove at a constant rate, how many miles did the car travel per hour?

ANSWER KEY
MODEL DRAWING
DECIMALS

Charlene bought 8 bowls and 8 plates for her new apartment. Each plate cost $.60 more than each bowl. Charlene spent $60.80 for the bowls and plates altogether. What was the cost of 1 bowl? What was the cost of 1 plate?

Charlene's bowls:

$3.50	$3.50	$3.50	$3.50	$3.50	$3.50	$3.50	$3.50
?							

? | $4.10

Charlene's plates:

$3.50	$3.50	$3.50	$3.50	$3.50	$3.50	$3.50	$3.50
+$.60	+$.60	+$.60	+$.60	+$.60	+$.60	+$.60	+$.60

$60.80

A. 8 X $.60 = $4.80

B. $60.80 - $4.80 = $56

C. 16 units = $56
$56 ÷ 16 = $3.50
1 unit = $3.50

D. $3.50 + $.60 = $4.10

Each bowl cost $3.50 and each plate cost $4.10.

ANSWER KEY
MODEL DRAWING
RATE

A car traveled 390 miles in 6 hours. If the car drove at a constant rate, how many miles did the car travel per hour?

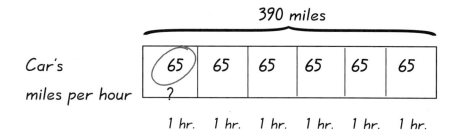

6 units = 390

390 ÷ 6 = 65

1 unit = 65

The car traveled 65 miles per hour.

The Parent Connection for Singapore Math

MODEL DRAWING

RATE

Mrs. Watson can type 70 words per minute. How many words can Mrs. Watson type in 1 hour?

RATE

The train traveled 280 miles in 4 hours. How far will the train travel in 7 hours?

ANSWER KEY
MODEL DRAWING

RATE

Mrs. Watson can type 70 words per minute. How many words can Mrs. Watson type in 1 hour?

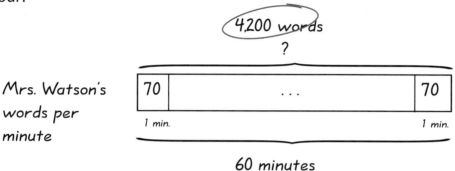

1 unit = 70

60 X 70 = 4,200

60 units = 4,200

Mrs. Watson can type 4,200 words in 60 minutes or 1 hour.

RATE

The train traveled 280 miles in 4 hours. How far will the train travel in 7 hours?

280 miles

Train's distance traveled

70	70	70	70	70	70	70

1 hr. 1 hr. 1 hr. 1 hr. 1 hr. 1 hr. 1 hr.

? 490 miles

A. 4 units = 280

280 ÷ 4 = 70

1 unit = 70

B. 1 unit = 70

7 X 70 = 490

7 units = 490

The train will travel 490 miles in 7 hours.

MODEL DRAWING

RATE

Bus A and Bus B covered the same distance of 350 miles. Bus A reached its destination in 5 hours. Bus B reached its destination in 7 hours. How many miles per hour was Bus A traveling? How many miles per hour was Bus B traveling?

RATIO

The ratio of girls to boys is 4 : 6. If there are 12 girls, how many boys are there? How many girls and boys are there altogether?

ANSWER KEY
MODEL DRAWING
RATE

   Bus A and Bus B covered the same distance of 350 miles. Bus A reached its destination in 5 hours. Bus B reached its destination in 7 hours. How many miles per hour was Bus A traveling? How many miles per hour was Bus B traveling?

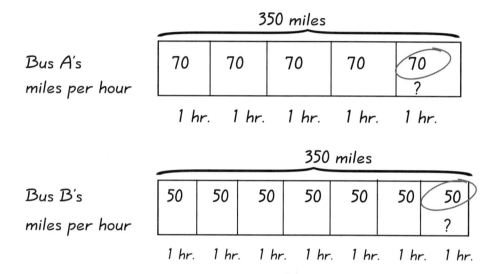

A. 5 units = 350

   350 ÷ 5 = 70

   1 unit = 70

B. 7 units = 350

   350 ÷ 7 = 50

   1 unit = 50

Bus A was traveling 70 miles per hour. Bus B was traveling 50 miles per hour.

ANSWER KEY
MODEL DRAWING

RATIO

The ratio of girls to boys is 4 : 6. If there are 12 girls, how many boys are there? How many girls and boys are there altogether?

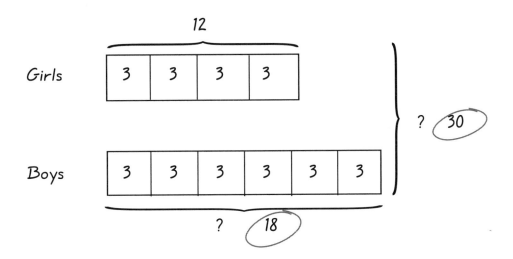

A. 4 units = 12

   12 ÷ 4 = 3

   1 unit = 3

B. 6 X 3 = 18

C. 12 + 18 = 30

There are 18 boys and 30 girls and boys altogether.

MODEL DRAWING

RATIO

The ratio of Jana's money to Karen's money is 4 : 3. If Jana has $120, how much money does Karen have? How much money do they have altogether?

RATIO

The ratio of knives to spoons to forks is 3 : 4 : 5. If there are 15 forks, how many spoons are there? How many utensils are there altogether?

ANSWER KEY
MODEL DRAWING
RATIO

The ratio of Jana's money to Karen's money is 4 : 3. If Jana has $120, how much money does Karen have? How much money do they have altogether?

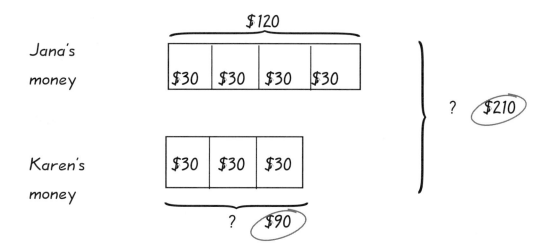

A. 4 units = $120

    $120 ÷ 4 = $30

    1 unit = $30

B. 3 X $30 = $90

C. $120 + $90 = $210

Karen has $90. Jana and Karen have $210 altogether.

ANSWER KEY
MODEL DRAWING
RATIO

The ratio of knives to spoons to forks is 3 : 4 : 5. If there are 15 forks, how many spoons are there? How many utensils are there altogether?

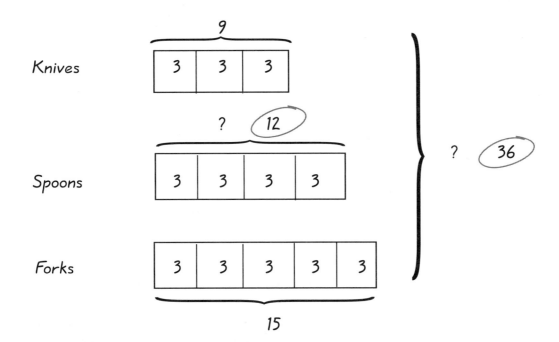

A. 5 units = 15
   15 ÷ 5 = 3
   1 unit = 3

B. 3 X 3 = 9

C. 4 X 3 = 12

D. 9 + 12 + 15 = 36

There are 12 spoons, and there are 36 utensils altogether.

MODEL DRAWING
PERCENT

At a movie theater, 30% of the seats were occupied by adults, 40% were occupied by children, and the rest were empty. If there were 120 seats in the theater, how many children were at the theater?

PERCENT

Of last year's graduating class at Rutgers, 75% of the students found jobs within the first month of graduating. If there were 4,500 graduating students, how many students found jobs within the first month after graduating?

ANSWER KEY
MODEL DRAWING
PERCENT

At a movie theater, 30% of the seats were occupied by adults, 40% were occupied by children, and the rest were empty. If there were 120 seats in the theater, how many children were at the theater?

A. 10 units = 120

   120 ÷ 10 = 12

   1 unit = 12

B. 1 unit = 12

   4 X 12 = 48

   4 units = 48

There were 48 children at the movie theater.

ANSWER KEY
MODEL DRAWING
PERCENT

Of last year's graduating class at Rutgers, 75% of the students found jobs within the first month of graduating. If there were 4,500 graduating students, how many students found jobs within the first month after graduating?

A. 4 units = 4,500

    4,500 ÷ 4 = 1,125

    1 unit = 1,125

B. 1 unit = 1,125

    3 X 1,125 = 3,375

    3 units = 3,375

In last year's Rutgers graduating class, 3,375 students found jobs within the first month after graduating.

MODEL DRAWING

PRE-ALGEBRA

The sum of 2 consecutive even numbers is 126. What are the 2 numbers?

● ●

PRE-ALGEBRA

Marta is now 3 times as old as Carlos. If Marta was 20 years old 4 years ago, how old is Carlos now?

ANSWER KEY
MODEL DRAWING

PRE-ALGEBRA

The sum of 2 consecutive even numbers is 126. What are the 2 numbers?

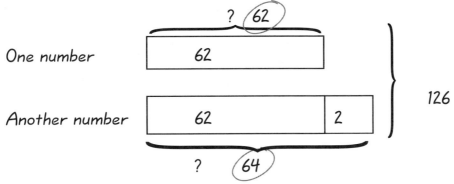

A. 126 – 2 = 124

B. 2 units = 124
    124 ÷ 2 = 62
    1 unit = 62

One number is 62 and the other number is 64.

PRE-ALGEBRA

Marta is now 3 times as old as Carlos. If Marta was 20 years old 4 years ago, how old is Carlos now?

24

Marta's
age now

| 8 | 8 | 8 |

3 units = 24
24 ÷ 3 = 8
1 unit = 8

Carlos's
age now

8
?

Carlos is now 8 years old.

PLACE VALUE DISKS

(1) (1) (1) (1)

(1) (1) (1) (1)

(1) (1) (1) (1)

(1) (1) (1) (1)

(1) (1) (1) (1)

PLACE VALUE DISKS

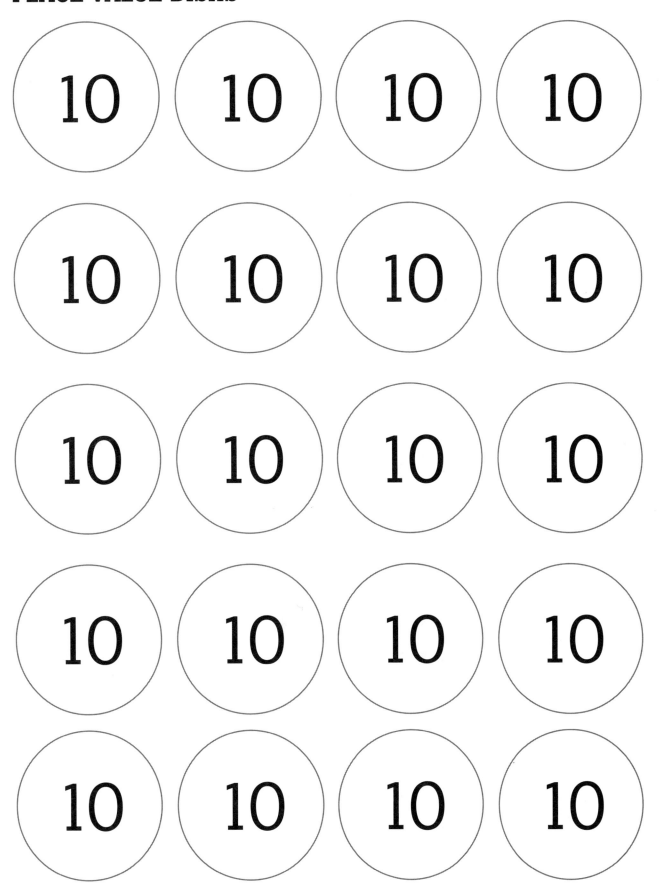

PLACE VALUE DISKS

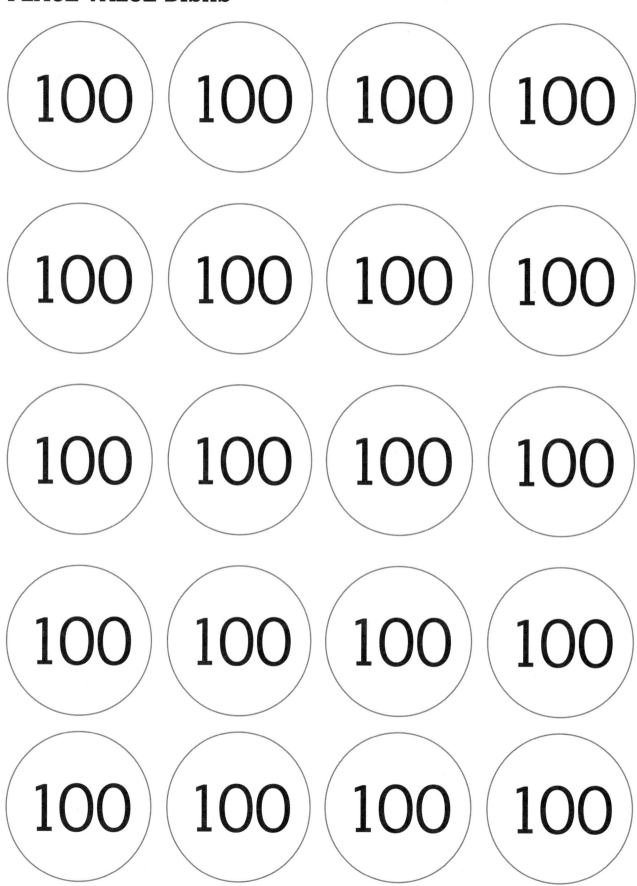

PLACE VALUE DISKS

1,000	1,000	1,000	1,000
1,000	1,000	1,000	1,000
1,000	1,000	1,000	1,000
1,000	1,000	1,000	1,000
1,000	1,000	1,000	1,000

PLACE VALUE DISKS

10,000	10,000	10,000	10,000
10,000	10,000	10,000	10,000
10,000	10,000	10,000	10,000
10,000	10,000	10,000	10,000
10,000	10,000	10,000	10,000

PLACE VALUE DISKS

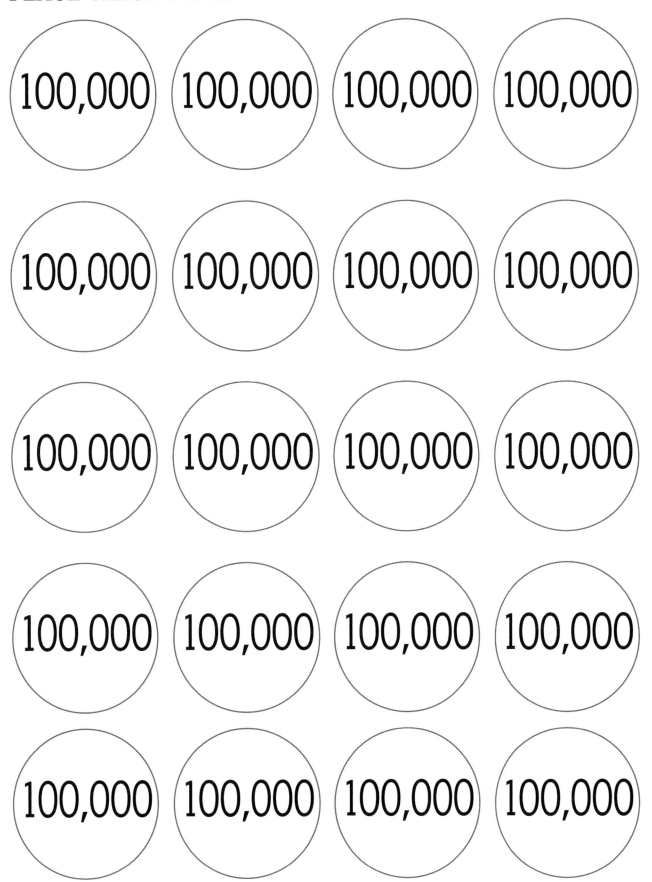

PLACE VALUE DISKS

1,000,000	1,000,000	1,000,000	1,000,000
1,000,000	1,000,000	1,000,000	1,000,000
1,000,000	1,000,000	1,000,000	1,000,000
1,000,000	1,000,000	1,000,000	1,000,000
1,000,000	1,000,000	1,000,000	1,000,000

HEADERS FOR PLACE VALUE MATS

Millions

Hundreds

Hundred Thousands

Tens

Ten Thousands

Ones

Thousands

PLACE VALUE STRIPS

1,0 0 0

2,0 0 0

3,0 0 0

4,0 0 0

The Parent Connection for Singapore Math

PLACE VALUE STRIPS

5,0 0 0

6,0 0 0

7,0 0 0

8,0 0 0

PLACE VALUE STRIPS

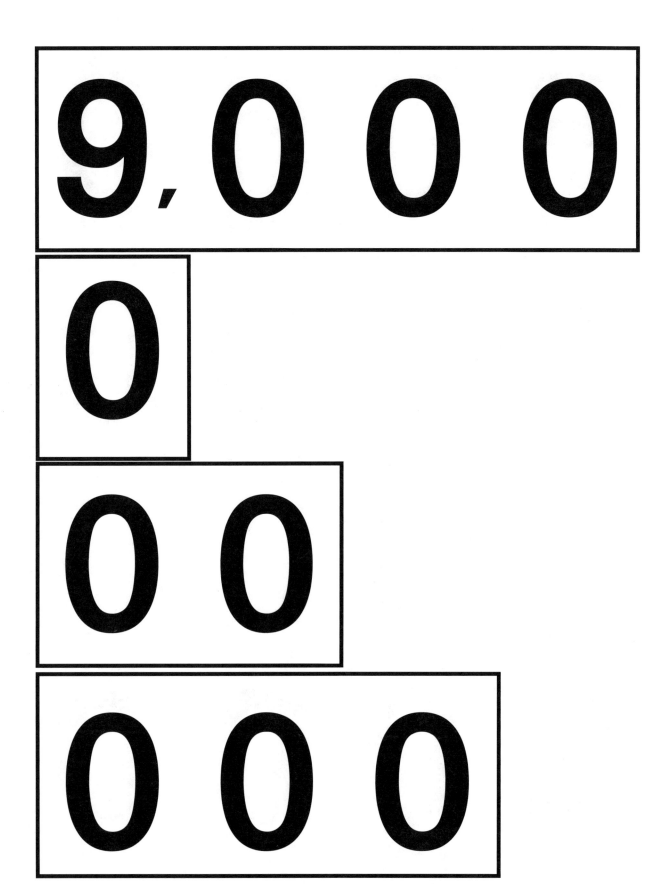

PLACE VALUE STRIPS

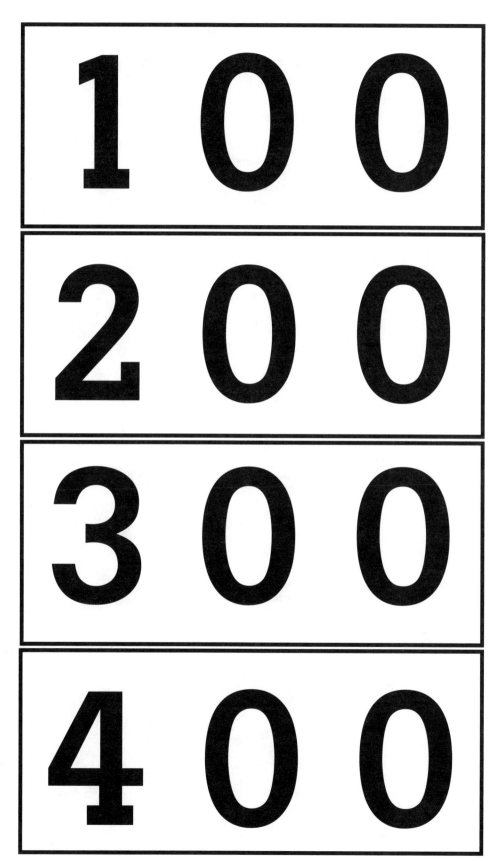

PLACE VALUE STRIPS

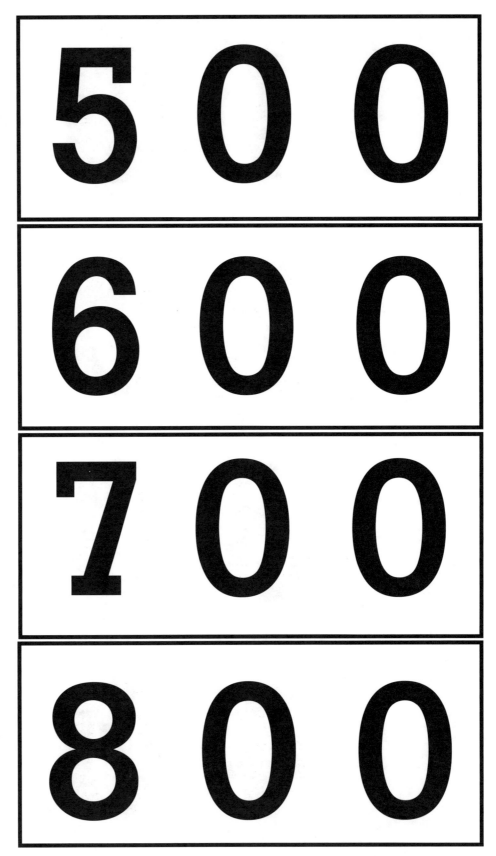

PLACE VALUE STRIPS

900

10

40

20

50

30

60

PLACE VALUE STRIPS

7 0   1

8 0   2   3

9 0   4   5

6   7   8   9

PLACE VALUE MAT

Ten Thousands	Thousands	Hundreds	Tens	Ones

PROBLEMS FOR PLACE-VALUE PRACTICE

ADDITION WITHOUT REBUNDLING/TRADING

3 + 4

2 + 6

4 + 5

ADDITION WITH REBUNDLING/TRADING

6 + 9

8 + 5

12 + 19

36 + 54

192 + 489

SUBTRACTION WITHOUT REBUNDLING/BORROWING

9 − 5

7 − 2

SUBTRACTION WITH REBUNDLING/BORROWING

13 − 8

25 − 16

41 − 23

374 − 289

PROBLEMS FOR PLACE-VALUE PRACTICE

MULTIPLICATION

3×4

6×9

23×7

239×8

DIVISION

$46 \div 2$

$124 \div 4$

$233 \div 7$

$641 \div 5$

NUMBER BONDS

- Focus on understanding the bond among 3 numbers
- Reduce the number of "math facts" that need to be learned

EXAMPLE:

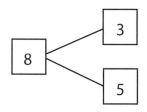

3 + 5 = 8

5 + 3 = 8

8 − 3 = 5

8 − 5 = 3

One number bond takes the place of 4 math facts!

The same principle can be applied to multiplication and division.

COMPUTATION STRATEGY PRACTICE
BRANCHING—ADDITION

Follow these steps to solve addition problems.

1. Write the problem horizontally.

2. Branch the numbers in the two addends into tens and ones, placing the tens on the outside and the ones on the inside.

3. Circle all the tens.

4. Continue to branch if necessary to make combinations of ten.

5. Compute.

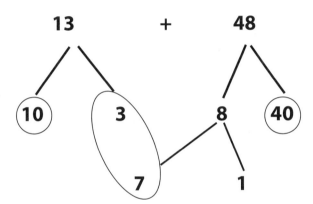

10 + 10 + 40 = 60

60 + 1 = 61

PRACTICE PROBLEMS

14 + 15

12 + 11

32 + 29

77 + 48

ANSWER KEY
BRANCHING—ADDITION

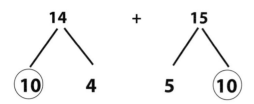

$10 + 10 = 20$

$4 + 5 = 9$

$20 + 9 = 29$

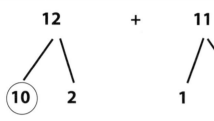

$10 + 10 = 20$

$2 + 1 = 3$

$20 + 3 = 23$

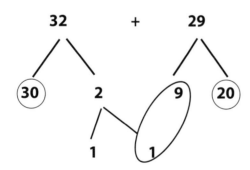

$30 + 10 + 20 = 60$

$60 + 1 = 61$

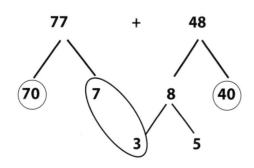

$70 + 10 + 40 = 120$

$120 + 5 = 125$

COMPUTATION STRATEGY PRACTICE
BRANCHING—SUBTRACTION

Follow these steps to solve subtraction problems.

1. Set up the problem horizontally.

2. The goal is to create bonds of ten. To do that, branch the numbers into tens and ones, placing the tens on the outside and the ones on the inside.

3. If necessary, rebundle (borrow) from the outside to the middle.

4. Circle the multiples of ten.

5. Subtract the numbers on the outside (tens).

6. Subtract the numbers on the inside (ones).

7. Add together the differences from the tens and the ones.

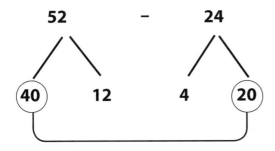

(40 − 20) + (12 − 4)

20 + 8 = 28

PRACTICE PROBLEMS

19 − 12

57 − 33

82 − 45

77 − 29

ANSWER KEY
BRANCHING—SUBTRACTION

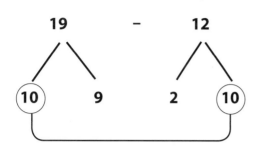

10 – 10 = 0

9 – 2 = 7

0 + 7 = 7

Remember to bring it all together in the end!

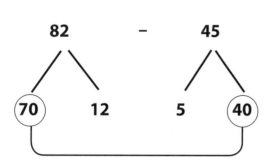

50 – 30 = 20

7 – 3 = 4

20 + 4 = 24

Remember to bring it all together in the end!

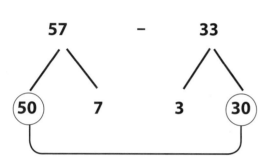

70 – 40 = 30

12 – 5 = 7

30 + 7 = 37

Remember to bring it all together in the end!

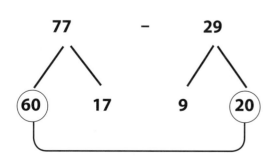

60 – 20 = 40

17 – 9 = 8

40 + 8 = 48

Remember to bring it all together in the end!

COMPUTATION STRATEGY PRACTICE
AREA MODEL FOR MULTIPLICATION

- Used for 2-digit by 2-digit multiplication and up
- Focuses on understanding the value of digits

EXAMPLE: 23 X 63

1. Break numbers into expanded form.

$23 = 20 + 3$ $63 = 60 + 3$

2. Draw a square and cut it in half both horizontally and vertically.

3. Distribute the expanded-form numbers across the top and along the left side.

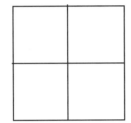

4. Multiply each pair of numbers, placing the product in the corresponding box.

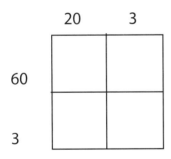

$20 \times 60 = 1{,}200$

$3 \times 60 = 180$

$20 \times 3 = 60$

$3 \times 3 = 9$

5. Add up all of the numbers in the boxes.

$$1,200 + 180 + 60 + 9 = 1,449$$

6. The sum of the addition problem becomes the product of your original multiplication problem.

$$23 \times 63 = 1,449$$

Try to incorporate mental-math strategies by completing the multiplication problems and addition problems without pencil and paper!

PRACTICE PROBLEMS

37×45

652×857

ANSWER KEY
AREA MODEL FOR MULTIPLICATION

37 X 45

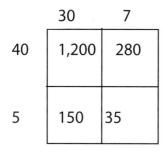

$30 \times 40 = 1{,}200$
$7 \times 40 = 280$
$30 \times 5 = 150$
$7 \times 5 = 35$

$1{,}200 + 280 + 150 + 35 = 1{,}665$
$37 \times 45 = 1{,}665$

652 X 857

	600	50	2
800	480,000	40,000	1,600
50	30,000	2,500	100
7	4,200	350	14

$480{,}000 + 40{,}000 + 1{,}600 = 521{,}600$

$30{,}000 + 2{,}500 + 100 = 32{,}600$

$4{,}200 + 350 + 14 = 4{,}564$

$521{,}600 + 32{,}600 + 4{,}564 = 558{,}764$

$652 \times 857 = 558{,}764$

COMPUTATION STRATEGY PRACTICE
VISUAL MODEL FOR ADDING FRACTIONS WITH UNLIKE DENOMINATORS

EXAMPLE: 1/5 + 2/3

1. Draw a large rectangle and create the first fraction by sectioning off and shading in the appropriate piece(s).

1/5

2. Begin to create the second fraction by drawing your lines in the opposite direction. Do not shade in the pieces yet.

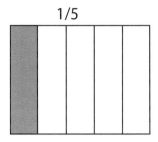

3. Before shading in your second fraction, discuss why we might not shade in the same pieces twice. (An analogy would be coloring a picture.)

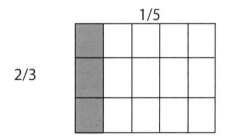

4. Count up the number of individual smaller pieces that would need to be shaded in order to represent your second fraction. (In this case, it would be 10 individual pieces.)

5. Shade in those pieces.

1/5

2/3
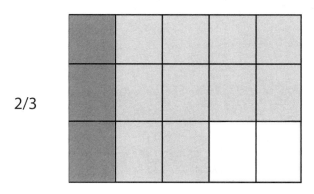

6. Count up your shaded parts for your numerator.

7. Count up your total parts for your denominator.

Final answer:

Shaded pieces =   13
Total pieces =   15

ANSWER KEY

VISUAL MODEL FOR ADDING FRACTIONS WITH UNLIKE DENOMINATORS

2/3

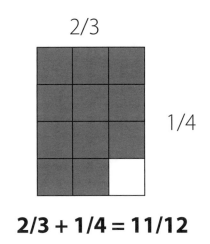

1/4

2/3 + 1/4 = 11/12

1/7

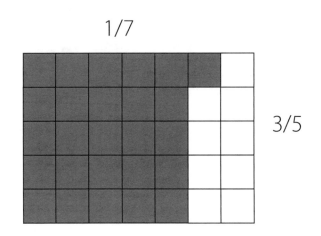

3/5

1/7 + 3/5 = 26/35

1/5

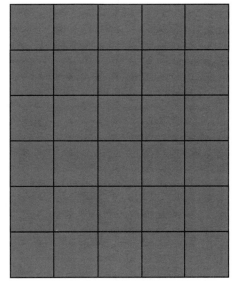

5/6

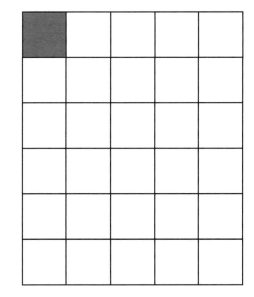

1/5 + 5/6 = 31/30   **OR**   **1 1/30**

COMPUTATION STRATEGY PRACTICE
QUESTIONING SEQUENCE FOR LEFT-TO-RIGHT COMPUTATION

EXAMPLE: *347 + 485*

1. When we add left to right, which place value do we start with? ***The place value farthest to the left.***

2. When adding 347 + 485, which numbers do we begin with? ***300 + 400.***

3. What is 300 + 400? ***700.***

4. Keep 700 in your head and move on to the next place value, the tens.

5. What will we be adding here? ***40 + 80.***

6. What is 40 + 80? ***120.***

7. What number do you currently have in your head? ***700.***

8. What is 700 + 120? ***820.***

9. Keep 820 in your head and move on to the next place value, the ones.

10. What will you be adding here? ***7 + 5.***

11. What is 7 + 5? ***12.***

12. What number are you holding in your head? ***820.***

13. What is 820 + 12? ***832.***

COMPUTATION STRATEGY PRACTICE

QUESTIONING SEQUENCE FOR REARRANGING

EXAMPLE: *637 X 5*

1. When multiplying using rearranging, which place value do we start with? ***The largest place value—in this case, the hundreds.***

2. What will we be multiplying first? ***600 X 5*** .

3. What is 600 X 5? ***3,000***.

4. Keep 3,000 in your head and move to the next place value, the tens.

5. What will you be multiplying? ***30 X 5***.

6. What is 30 X 5? ***150***.

7. What number are you holding in your head? ***3,000***.

8. What is 3,000 + 150? ***3,150***.

9. Keep 3,150 in your head and move to the next place value, the ones.

10. What will you be multiplying? ***7 X 5***.

11. What is 7 X 5? ***35***.

12. What number are you holding in your head? ***3,150***.

13. What is 3,150 + 35? ***3,185***.

PARENT INFORMATION NIGHT

INTRODUCTION TO SINGAPORE MATH

Singapore Math is a mastery-based program that incorporates many exciting strategies into the daily math routine. These strategies give students the opportunity to master concepts while gaining an in-depth understanding of numbers. Come join us to learn more about the hands-on program your student will be participating in during this part of his or her educational career.

When: _____

Where: _____

So we can be sure to have enough materials available, please return the bottom portion of this form to your student's teacher by _____.

. . . .

Parent Information Night
Introduction to Singapore Math

Student Name: _____

Grade: _____

Teacher: _____

PARENT INFORMATION NIGHT
MODEL DRAWING

Model drawing is one of the key components of the Singapore Math program. This approach to problem solving allows students to create a more concrete representation of the information in order to help them solve the problem. Join us for a fun night filled with exciting and challenging word problems.

Note: If you weren't able to make it to an earlier Singapore Math parent night, please don't let that keep you away. The parent night programs don't depend on each other, and we'd still love to have you join us for this one!

When: _____

Where: _____

So we can be sure to have enough materials available, please return the bottom portion of this form to your student's teacher by _____.

· ·

Parent Information Night
Model Drawing

Student Name: _____

Grade: _____

Teacher: _____

PARENT MAKE-&-TAKE NIGHT
PLACE VALUE DISKS & MATS

Throughout the year, your child will be using many manipulatives during math class. These tools help reinforce an understanding of place value, computation, fractions, decimals, geometry, measurement, and much more. As part of the Singapore Math program, we also work with some *new* manipulatives that help students reinforce many skills. Come join us to create a set of place value disks and mats. These can be used at home to show a wide range of mathematical concepts, and your child will have a blast showing you all of the exciting math we're doing in school!

Note: If you weren't able to make it to earlier Singapore Math parent nights, please don't let that keep you away. The parent night programs don't depend on each other, and we'd still love to have you join us for this one!

When: _____

Where: _____

So we can be sure to have enough materials available, please return the bottom portion of this form to your student's teacher by _____.

. . . .

Parent Make-&-Take Night
Place Value Disks & Mats

Student Name: _____

Grade: _____

Teacher: _____

The Parent Connection for Singapore Math 121

PARENT MAKE-&-TAKE NIGHT
PLACE VALUE STRIPS

As part of the Singapore Math program, we're using some new manipulatives this year to help students reinforce many skills. Come join us to make place value strips! These strips help to build an understanding of place value and the value of digits, and they can be used for computational strategies as well. Your child will have a blast showing you how these tools are used in school!

Note: If you weren't able to make it to earlier Singapore Math parent nights, please don't let that keep you away. The parent night programs don't depend on each other, and we'd still love to have you join us for this one!

When: _____

Where: _____

So we can be sure to have enough materials available, please return the bottom portion of this form to your student's teacher by _____.

. . . .

Parent Make-&-Take Night
Place Value Strips

Student Name: _____

Grade: _____

Teacher: _____

PARENT INFORMATION NIGHT

UNDERSTANDING PLACE VALUE

The concept of numbers can be very abstract. Many students learn the rules and follow the steps to compute mathematical equations, but not many of them understand what they are actually doing. Gaining an understanding of place value is key to developing the fundamentals of number sense. Join us for an exciting, hands-on evening devoted to place value and number sense!

If you were able to attend the Place Value Disks & Mats Make-&-Take or the Place Value Strips Make-& Take, please feel free to bring in your manipulatives. Otherwise, we'll be happy to provide manipulatives for you to use.

Note: If you weren't able to make it to earlier Singapore Math parent nights, please don't let that keep you away. The parent night programs don't depend on each other, and we'd still love to have you join us for this one!

When: _____

Where: _____

So we can be sure to have enough materials available, please return the bottom portion of this form to your student's teacher by _____.

· ·

Parent Information Night
Understanding Place Value

Student Name: _____

Grade: _____

Teacher: _____

PARENT INFORMATION NIGHT
COMPUTATION STRATEGIES

Once students have developed strong number sense, it's important to teach them computational strategies that require them to continue to call on that number sense. In class, we will be exploring many computational strategies—including branching, the area model for multiplication, and the visual model for adding fractions with unlike denominators, to name just a few. By joining us this evening, you'll learn how you can better help your student at home.

Note: If you weren't able to make it to earlier Singapore Math parent nights, please don't let that keep you away. The parent night programs don't depend on each other, and we'd still love to have you join us for this one!

When: _____

Where: _____

So we can be sure to have enough materials available, please return the bottom portion of this form to your student's teacher by _____.

· ·

Parent Information Night
Computation Strategies

Student Name: _____

Grade: _____

Teacher: _____

Index

A

Addition
    branching strategy for, 16, 31, 107–8
    of fractions (*see* Visual model for adding fractions with unlike denominators)
    model drawing for, 44, 48, 50–53
    problems for place-value practice, 104
Area model for multiplication
    as back-to-school-night icebreaker, 13–14, 111–13
    explaining, at Computation Strategies parent night, 31, 111–13
    send-home on, 17, 111–13

B

Back-to-school night
    icebreakers for
        area model for multiplication, 13–14, 111–13
        branching, 13, 107–8
        grading a model-drawing solution, 11–12, 37–43, 50–87
        mental math, 12, 117–18
        model drawing, 11, 37–43, 50–87
        place value strips, 14, 96–102
        purpose of, 6, 10
    parent reaction to, 10
    timing of, 6
Branching
    addition, 16, 31, 107–8
    as back-to-school-night icebreaker, 13, 107–8
    explaining, at Computation Strategies parent night, 31, 107–10
    send-home on, 16, 107–10
    subtraction, 16, 31, 109–10

C

Class structure, for Singapore Math, 36
Computation Strategies parent night, 19, 31–32, 106–18
    notice and response slip for, 124

D

Decimals, model drawing for, 69, 71–72, 73
Denominators, unlike. *See* Fractions, with unlike denominators
Division
    model drawing for, 62–65
    problems for place-value practice, 105

F

Fractions
    model drawing for, 47, 49, 66–69, 70
    with unlike denominators, visual model for adding
        explaining, at Computation Strategies parent night, 32, 114–16
        send-home on, 17, 114–16

H

Handouts. *See also* Send-homes
    for non-English-speaking parents, 9
    preparing, for parent nights, 8, 20, 21
Headers for place value mats, 95

I

Icebreakers, back-to-school-night
    area model for multiplication, 13–14, 111–13
    branching, 13, 107–8
    grading a model-drawing solution, 11–12, 37–43, 50–87
    mental math, 12, 117–18
    model drawing, 11, 37–43, 50–87
    place value strips, 14, 96–102
    purpose of, 6, 10
Introduction to Singapore Math parent night, 18, 22–23, 34–38, 40, 42
    notice and response slip for, 119
    when to offer, 6, 7

K

Key reference points, on Singapore Math, 35

Q

Questioning sequence
    for left-to-right computation, 12, 117
    for rearranging, 118

R

Rate, model drawing for, 72, 74–77, 78
Ratio, model drawing for, 77, 79–82
Rearranging, 32, 118
Reference points on Singapore Math, key, 35
Reproducibles, how to use, 6
Response slips, for parent night notices, 20–21, 119–24
Rubrics, 11, 12, 22

S

Send-homes
    benefits of, 6
    helping parents understand, 8–9
    for non-English-speaking parents, 9
    parent receptivity to, 10
    purpose of, 6, 15
    topics for
        area model for multiplication, 17, 111–13
        branching, 16, 107–10
        choosing, 15
        model drawing, 16, 37, 44–49
        visual model for adding fractions with unlike denominators, 17, 114–16
    when to use, 6, 15
Singapore Math program
    benefits of, 7
    difficulty implementing, 5
    key reference points on, 35
    parent reactions to, 5, 6, 10
    strategies for parent involvement in (*see* Icebreakers, back-to-school-night; Parent nights; Send-homes)
        when to use, 7
Student achievement, as goal of Singapore Math, 7
Subtraction
    branching strategy for, 16, 31, 109–10
    model drawing for, 45, 48, 54–57
    problems for place-value practice, 104

T

TIMSS (Trends in International Mathematics & Science Study) report, 22, 34
Tutoring sessions, for explaining send-homes, 9

U

Understanding Place Value parent night, 19, 29–30, 103–5
    notice and response slip for, 123

V

Visual model for adding fractions with unlike denominators
    explaining, at Computation Strategies parent night, 32, 114–16
    send-home on, 17, 114–16

Also by Sandra Chen

Math Manipulatives

Place Value Decimal Tiles

Place Value Overhead Decimal Tiles

Place Value Decimal Strips

Place Value Decimal Cubes

CD-ROM Presentation

Singapore Math: Place Value, Computation & Number Sense

Bring Sandra Chen right to your school for on-site training! To learn how, call (877) 388-2054.